CW01521222

Brooks/Cole

One-Unit Series

in Precalculus Mathematics

# ANALYTIC GEOMETRY

## Second Edition

Brooks/Cole

One-Unit Series

in Precalculus Mathematics

# ANALYTIC GEOMETRY

## Second Edition

Karl J. Smith

BROOKS/COLE PUBLISHING COMPANY

Pacific Grove, California

Brooks/Cole Publishing Company
A Division of Wadsworth, Inc.

© 1991, 1975 by Wadsworth, Inc., Belmont, California 94002.   All rights reserved. This material may not be reproduced, stored in a retrieval system, or transcribed, in any form or by any means — electronic, mechanical, photocopying, or otherwise — without the prior written permission of the publisher: Brooks/Cole Publishing Company, Pacific Grove, California 93950, a division of Wadsworth, Inc.

Printed in the United States of America

10   9   8   7   6   5   4   3   2   1


Library of Congress Cataloging-in-Publication Data

Smith, Karl J.
     Analytic Geometry/ Karl J. Smith
          p.   cm.   --   (Brooks/Cole one-unit series in precalculus mathematics)
     Includes index.
     ISBN 0-534-14928-6  (paper)
     1. Geometry, Analytic.    I. Title    II. Series
  QA551.S614    1991
  516.3--dc20                                                                    90-24937
                                                                                      CIP

Sponsoring Editor:  Paula-Christy Heighton
Editorial Assistant:  Elaine Giuliano
Marketing Representative:  Deborah Mobley
Production Editor:  Joan Marsh
Production Service:  Bookman Productions
Manuscript Editor:  Steven Gray
Interior Design:  Karl J. Smith
Cover Design:  Vernon T. Boes
Typesetting:  Karlin Enterprises
Cover Printing:  Malloy Lithographing, Inc.
Printing and Binding:  Malloy Lithographing, Inc.

# PREFACE

For a number of years, I've noticed that many of my students taking beginning calculus lack sufficient preparation in analytic geometry. Even though most calculus books include analytic geometry, it often comes later in the book and assumes a great deal of knowledge about the principles of analytic geometry. This book is written to provide a review of those principles. The material can be used as a supplement to a trigonometry course, as a reference in a calculus course, or as the basis for a short (one unit) course.

Since this course is a refresher, I will not attempt to develop or prove all the results used in this book. Most of the problem sets consist of both A and B problems. The A problems deal with developing skill in analytic geometry, and the B problems deal with theory. A person who wishes to prove the results stated in the text should work the B problems. However, a careful working of the A problems will provide a firm foundation and review of analytic geometry. For this reason, the answers to most of the A problems are given in an appendix.

I would like to extend my thanks to the following reviewers for their comments and suggestions: Beth Hooper, Golden West College; Henry Tjoelker, California State University, Sacrament; and Carroll Wells, Western Kentucky University.

In particular, my appreciation to Carol Kublin for checking all of the examples and answers for accuracy.

Karl J. Smith
Sebastopol, CA

# CONTENTS

---

# CHAPTER 1    THE STRAIGHT LINE

### 1.1 Preliminaries

As stated in the preface, this text is designed as a refresher course in analytic geometry. I will therefore be making some assumptions about your previous background and will not attempt to develop or prove all the results used in this book. Instead, I will generally state the main ideas you will need for calculus in the text and ask for the proofs or justifications in the problem sets (B problems).

I will assume that you are familiar with the Cartesian coordinate system, with related terminology, and with the plotting of points, as summarized in Figure 1.

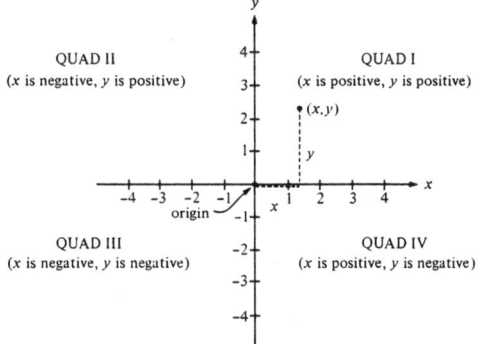

Figure 1    Cartesian coordinate system

In analytic geometry, we are concerned with the relationship between algebra and geometry. This idea occurred first to a French mathematician, René Descartes (1596–1650). It has been said that the original idea came to Descartes in a flash as he was lying in bed watching a fly crawl on the ceiling near a corner of his room. He saw that he could describe the path of the fly if he knew the relation connecting the fly's distances from the two adjacent walls. He was able to describe this path algebraically and thus linked together algebra and geometry.

The idea is a simple but very important one. In algebra we talk of sets of ordered pairs as relations, and in geometry we talk of sets of ordered pairs as curves. If we relate a set of ordered pairs of a relation to a set of ordered pairs of a curve in a one-to-one fashion, we have the *graph of an equation* or *the equation of a graph.*

**GRAPH OF AN
EQUATION OR
EQUATION OF A
GRAPH**

By the **graph of an equation** or the **equation of a graph** we mean:
There is a one-to-one correspondence between ordered pairs
$(x, y)$ that satisfy the equation and ordered pairs $(x, y)$ that lie
on the curve.

For example, all the pairs $(x, y)$ that satisfy the equation

$$(x^2 + y^2 - 2x)^2 = 4(x^2 + y^2)$$

are the same as the pairs $(x, y)$ that lie on the curve shown in Figure
2.

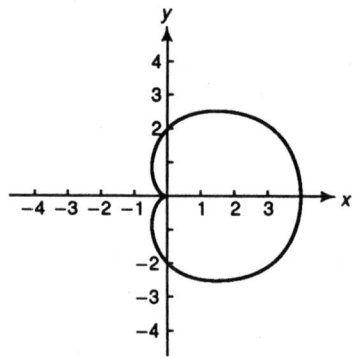

Figure 2　Graph of $(x^2 + y^2 - 2x)^2 = 4(x^2 + y^2)$

In analytic geometry we study this relationship between
equation and graph and arrive at two main problems:

1.　　Given an equation, find the graph.

2.　　Given a graph (or sometimes the conditions of a graph), find
the corresponding equation.

The method of development in this course will be to begin with
the simplest types of equations:　first-degree equations in two
variables.　We will then work our way through second- and higher-
degree equations, and finally we will consider polar-form equations.
We therefore begin with the line represented by an equation of the
first degree in two variables.

### 1.2　Distance Formula

**DISTANCE
FORMULA**

The distance $d$ between two points $P_1(x_1, y_1)$ and $P_2(x_2, y_2)$ is

$$d = \sqrt{(x_2 - x_1)^2 + (y_2 - y_1)^2}$$

Throughout the book we will use subscripts on $x$ and $y$
whenever they denote *known* or *given* points. Thus, $(x_1, y_1)$, $(x_2, y_2)$,
and $(x_3, y_3)$ denote known particular points, whereas $(x, y)$ denotes
*any* point satisfying some given conditions.

**EXAMPLE 1**

Find the distance between the given points.
**a.** $(10, 4)$ and $(15, -8)$
**b.** $(5, -3)$ and $(-1, 4)$
**c.** $(0, c)$ and $(x, y)$
**d.** $(4, 5)$ and the $x$-axis

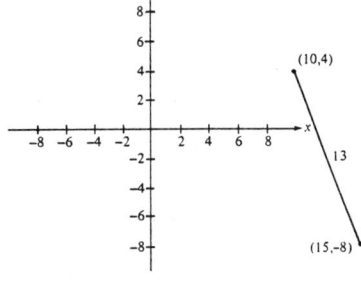

*Solution*

**a.**
$$d = \sqrt{(15 - 10)^2 + (-8 - 4)^2}$$
$$= \sqrt{25 + 144}$$
$$= 13$$

**b.**
$$d = \sqrt{(-1 - 5)^2 + (-4 + 3)^2}$$
$$= \sqrt{36 + 1}$$
$$= \sqrt{37}$$

**c.**
$$d = \sqrt{(x - 0)^2 + (y - c)^2}$$
$$= \sqrt{x^2 + y^2 - 2cy + c^2}$$

**d.** By the distance from a point to a line, we mean the perpendicular distance. Thus, we use the distance formula between $(4, 5)$ and $(4, 0)$:
$$d = \sqrt{(4 - 4)^2 + (5 - 0)^2}$$
$$= \sqrt{25}$$
$$= 5$$

We can also work this example more easily by noting that, by the definition, $(4, 5)$ means

4 units from the $y$-axis
5 units from the $x$-axis

This is shown in Figure 3.

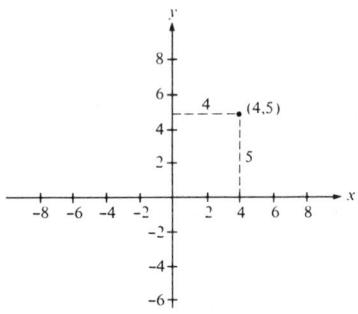

Figure 3  Distance of $(4, 5)$ from the coordinate axis     □

**EXAMPLE 2**

Find the value of $x_2$ if the distance from $(1, -1)$ to $(x_2, 2)$ is $3\sqrt{2}$. Notice that the point $(x_2, 2)$ lies on the line $y = 2$.

*Solution*

Let $(x_2, 2)$ be the coordinates of the point we wish to find. We see that $d = 3\sqrt{2}$, and thus we have

$$3\sqrt{2} = \sqrt{(x_2 - 1)^2 + (2 + 1)^2}$$

Squaring both sides (see Appendix A for solving quadratic equations), we obtain

$$9 \cdot 2 = (x_2 - 1)^2 + 9$$

$$18 = x_2{}^2 - 2x_2 + 1 + 9$$

$$x_2{}^2 - 2x_2 - 8 = 0$$

$$(x_2 + 2)(x_2 - 4) = 0$$

$$x_2 = 4, -2$$

We see that there are two possibilities, $(4, 2)$ and $(-2, 2)$. Both of these points are $3\sqrt{2}$ units from $(1, -1)$, as shown in Figure 4.

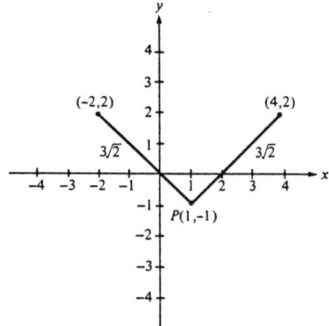

Figure 4   Sketch for Example 2                               □

### 1.3  Slope

Another property of lines is the *inclination of a line*, as illustrated in Figure 5.

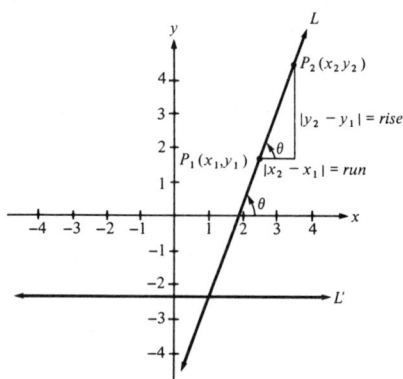

Figure 5   Angles of inclination

**INCLINATION OF
A LINE**

The **inclination of a line** $L$ that is not parallel to the $x$-axis is the smallest positive angle between the $x$-axis and line $L$. If $L$ is parallel to the $x$-axis, the inclination is defined to be zero.

We need the concept of the inclination of a line to be able to define the **slope** of a line. In calculus you will generalize this idea to include the slope of any curve.

**SLOPE OF A LINE**

The **slope of a line** $L$ is

$$m = \tan \theta = \frac{\text{RISE}}{\text{RUN}}$$

where $\theta$ is the angle of inclination of $L$.

*Notes:*

1. Using the definition of tan $\theta$, we see that, if $L$ passes through $P(x_1, y_1)$ and $P_2(x_2, y_2)$, then

$$m = \tan \theta = \frac{y_2 - y_1}{x_2 - x_1}$$

2. We see that $y_2 - y_1$ is the rise for a given run of $x_2 - x_1$. Thus, sometimes we say

$$m = \frac{\text{RISE}}{\text{RUN}}$$

3. The slope is undefined if $x_1 = x_2$ (in which case the line $L$ is vertical, i.e., $\theta = 90°$; notice also that tan $90°$ is undefined).

**EXAMPLE 3**

Find the slope of the line passing through the given points.

**a.** $(10, 4)$ and $(15, -8)$    **b.** $(4, 5)$ and $(6, 5)$

*Solution*

**a.**
$$m = \tan \theta$$
$$= \frac{-8 - 4}{2}$$
$$= \frac{-12}{5}$$

**b.**
$$m = \tan \theta$$
$$= \frac{5 - 5}{6 - 4}$$
$$= 0 \qquad \qquad \square$$

Notice in Example 3(b) that a horizontal line has an angle of inclination of $0°$. Since tan $0° = 0$, we could have noted the general result summarized in the following box.

**HORIZONTAL LINES**

Horizontal lines have zero slope.

**EXAMPLE 4**

Find the slope of the line passing though $(4, 5)$ and $(4, 1)$.

*Solution*

We see that the line is a vertical line, and thus the angle of inclination is $90°$ ($m = \tan 90°$ is undefined). $\qquad \square$

Example 4 leads us to yet another general result.

**VERTICAL LINES**

Vertical lines have no slope.

Notice that saying that a line has no slope is not the same as saying that the line has zero slope.

### 1.4  Parallel and Perpendicular Lines

**PARALLEL LINES**

**Theorem 1:** Two nonvertical lines are parallel if and only if their slopes are equal.

**PERPENDICULAR LINES**

**Theorem 2:** Two nonvertical lines are perpendicular if and only if the product of their slopes is $-1$. Vertical and horizontal lines are perpendicular.

Another way of stating these theorems is to let $L_1$ *and* $L_2$ be lines with slopes $m_1$ and $m_2$, respectively. Then,

$$L_1 \parallel L_2 \text{ if and only if } m_1 = m_2$$

and

$$L_1 \perp L_2 \text{ if and only if } m_1 m_2 = -1$$

**EXAMPLE 5**

The points $A(4, -2)$, $B(8, 3)$ and $C(3, 7)$ are vertices of a triangle. Prove that $\triangle ABC$ is a right triangle.

*Solution*

$$\text{Slope of } AB = \frac{3 - (-2)}{8 - 4} = \frac{5}{4} = m_1$$

$$\text{Slope of } BC = \frac{7 - 3}{3 - 8} = \frac{4}{-5} = m_2$$

Since $m_1 m_2 = -1$, the line segments are perpendicular and $\triangle ABC$ is a right triangle. □

**EXAMPLE 6**

For $\triangle ABC$ in Example 5, show that the line through the midpoints of sides $AB$ and $AC$ is parallel to side $BC$ of the triangle.

*Solution*

Let $M_1$ and $M_2$ be the midpoints of $AB$ and $AC$, respectively, as shown in Figure 6. The midpoints of segments $P_1(x_1, y_1)$ and $P_2(x_2, y_2)$ are found by finding the average of the first components and the average of the second components. Thus,

$$M_1 = \left( \frac{8 + 4}{2}, \frac{3 + (-2)}{2} \right) = \left( 6, \tfrac{1}{2} \right)$$

$$M_2 = \left( \frac{3 + 4}{2}, \frac{7 + (-2)}{2} \right) = \left( \tfrac{7}{2}, \tfrac{1}{2} \right)$$

$$\text{Slope of } BC = -\tfrac{4}{5} \qquad \textit{from Example 5}$$

$$\text{Slope of } M_1 M_2 = \frac{\tfrac{5}{2} - \tfrac{1}{2}}{\tfrac{7}{2} - 6} = \frac{\tfrac{4}{2}}{-\tfrac{5}{2}} = \tfrac{4}{2} \cdot \left( -\tfrac{2}{5} \right) = -\tfrac{4}{5}$$

Since the slopes of $BC$ and $M_1 M_2$ are equal, the line segments $M_1 M_2$ and $BC$ are parallel. □

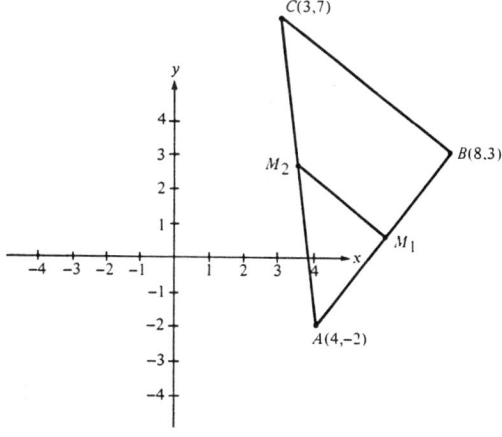

Figure 6 Sketch for Examples 5 and 6

The result noted in Example 6 is important enough to be restated as a general formula.

**MIDPOINT FORMULA**

> The midpoint $M$ of the segment connecting $P_1(x_1, y_1)$ and $P_2(x_2, y_2)$ is
> $$M = \left( \frac{x_1 + x_2}{2}, \frac{y_1 + y_2}{2} \right)$$

## 1.5 Equations of a Line

There are many ways of characterizing a line algebraically, so there are several forms of the equation of a line. (Remember that when we speak of "the equation of a line," we mean that a one-to-one correspondence must exist between points on the line and ordered pairs satisfying the equation.) These forms are summarized below.

**FORMS OF THE EQUATIONS OF A LINE**

> **Standard Form:** $Ax + By + C = 0$
> $(x, y)$ is any point on the line; $A$, $B$, *and* $C$ are constants
>
> **Slope–Intercept Form:** $y = mx + b$
> $m$ is the slope; $b$ is the $y$-intercept
>
> **Point–Slope Form:** $y - k = m(x - h)$
> $(x_1, y_1)$ is a given point
>
> **Two–point Form:** $y - y_1 = \left( \frac{y_2 - y_1}{x_2 - x_1} \right)(x - x_1)$
> $(x_1, y_1)$ and $(x_2, y_2)$ are the given points; $x_1 \neq x_2$

*In Examples 7–11, write the equations for the described lines in standard form, and graph each line.*

**EXAMPLE 7**

Passing through $(4, -3)$ with slope $\frac{2}{3}$.

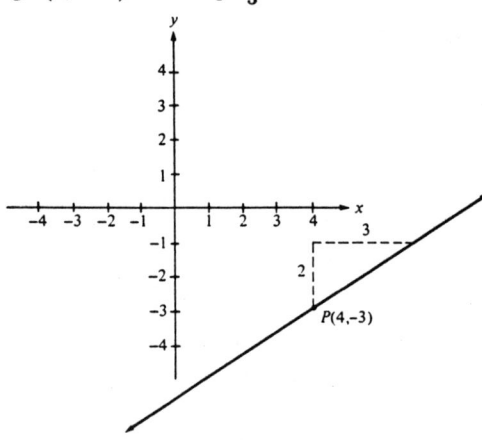

*Solution*

A point and the slope of the line are given, so you should use the point–slope form $y - y_1 = m(x - x_1)$.

$y + 3 = \frac{2}{3}(x - 4)$    *This is the equation of the line. We must now put it into standard form.*

$3y + 9 = 2(x - 4)$    *It is customary to eliminate fractions when writing an equation in standard form.*

$3y + 9 = 2x - 8$

$2x - 3y - 17 = 0$    *It is customary to write the standard form equation of a line with the leading coefficient positive.*    □

**EXAMPLE 8**

Slope $-\frac{5}{6}$ and $y$-intercept 7.

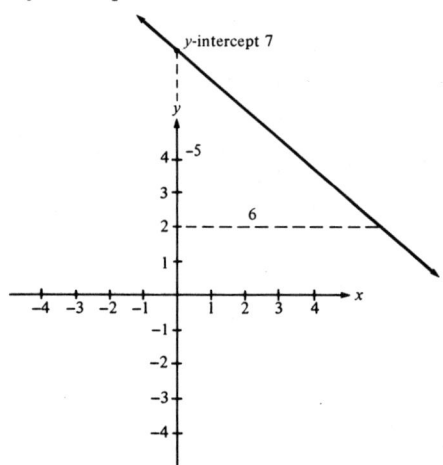

*Solution*    Use the slope–intercept form $y = mx + b$.

$$y = -\tfrac{5}{6}x + 7$$
$$6y = -5x + 42$$
$$5x + 6y - 42 = 0 \qquad \square$$

**EXAMPLE 9**    Passing through the points $(2, 3)$ and $(-4, -1)$.

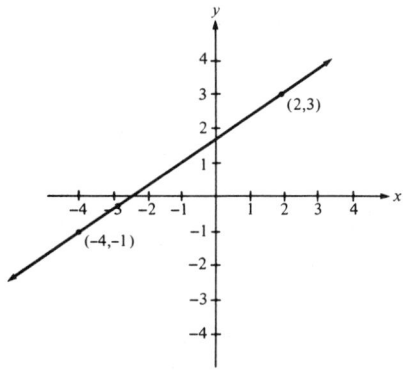

*Solution*    Use the two-point form $y - y_1 = \left(\dfrac{y_2 - y_1}{x_2 - x_1}\right)(x - x_1)$.

$$y - 3 = \left(\frac{-1 - 3}{-4 - 2}\right)(x - 2)$$
$$y - 3 = \tfrac{2}{3}(x - 2)$$
$$3y - 9 = 2x - 4$$
$$2x - 3y + 5 = 0 \qquad \square$$

**EXAMPLE 10**    Slope $\tfrac{3}{4}$ with $x$-intercept 5.

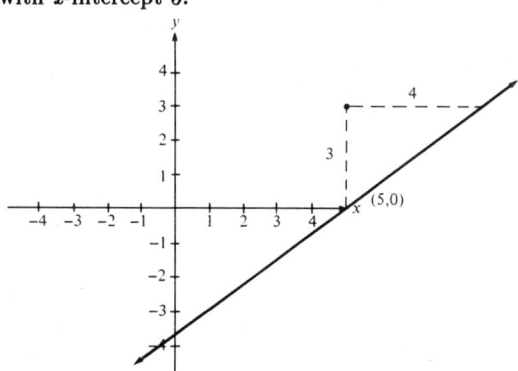

*Solution*    Use the point–slope form.  Be careful not to use slope–intercept form, since we are not given the $y$-intercept.

$$y - 0 = \tfrac{3}{4}(x - 15)$$
$$4y = 3x - 15$$
$$3x - 4y - 15 = 0 \qquad \square$$

**EXAMPLE 11**    $x$-intercept 3 and $y$-intercept $-5$.

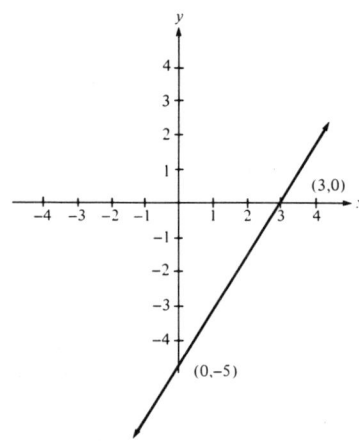

*Solution*    Use the two-point form for the points $(3, 0)$ and $(0, -5)$.

$$y - 0 = \left(\frac{-5 - 0}{0 - 3}\right)(x - 3)$$

$$y = \tfrac{5}{3}(x - 3)$$

$$3y = 5x - 15$$

$$5x - 3y - 15 = 0 \qquad\qquad \square$$

**EXAMPLE 12**    Graph $4x + y + 3 = 0$.

*Solution*    Put the equation into slope–intercept form by solving for $y$.

$$y = -4x - 3$$

By inspection, $m = -4$ and the $y$–intercept is $-3$. We now complete the graph as shown in Figure 7.

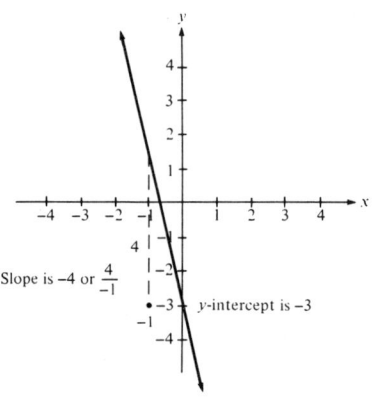

Figure 7   Graph of $4x + y + 3 = 0$ $\qquad\qquad \square$

**EXAMPLE 13**

*Solution*

Graph $y = 5$.

Lines whose equations are of the form $y = constant$ are horizontal lines. The equation $y = 5$ is simply a form of

$$Ax + By + C = 0 \qquad \text{where } A = 0, \ B = 1, \text{ and } C = -5:$$

$$0 \cdot x + 1 \cdot y - 5 = 0$$

The graph is shown in Figure 8.

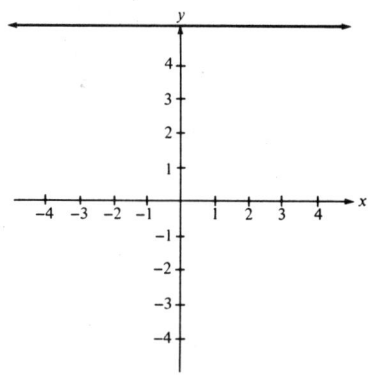

Figure 8 Graph of $y = 5$ □

**EXAMPLE 14**

*Solution*

Graph $x = -3$.

Lines whose equations are of the form $x = constant$ are vertical lines. The line $x = -3$ is shown in Figure 9.

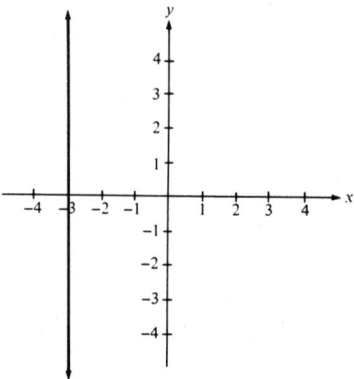

Figure 9 Graph of $x = -3$ □

## 1.6 PROBLEM SET 1

**A**

1. **a.** Draw the triangle whose vertices are $A(-2, -2)$, $B(3, 3)$, and $C(-1, 7)$.
   **b.** Find the perimeter of $\triangle ABC$.

   **c.** Using distances, show that the triangle is a right triangle.
   **d.** Using slopes, show that the triangle is a right triangle.
   **e.** Show that the line segment connecting the midpoints of $AB$ and $AC$ is parallel to side $BC$ of the triangle.

**2.**    **a.** Draw the quadrilateral whose vertices are $A(2, 2)$, $B(6, 4)$, $C(4, 8)$, and $D(0, 6)$.
   **b.** Find the perimeter of $ABCD$.
   **c.** Show that the quadrilateral is a parallelogram.
   **d.** Is the quadrilateral a rectangle?
   **e.** Show that the diagonals are perpendicular.

*In Problems 3–20, find the equation of the line in standard form, and sketch the graph.*

**3.**    Slope $-4$ passing through $(4, -2)$

**4.**    Passing through $(1, 5)$ with slope $-\frac{3}{4}$

**5.**    Slope $0$ passing through $(-2, -3)$

**6.**    No slope passing through $(-2, -3)$

**7.**    Slope $\frac{1}{6}$ with $y$-intercept $-4$

**8.**    Slope $-\frac{2}{5}$ with $y$-intercept $3$

**9.**    Slope $\frac{4}{3}$ with $x$-intercept $2$

**10.**   Slope $\frac{5}{2}$ with $x$-intercept $-3$

**11.**   Passing through the points $(1, 4)$ and $(2, 3)$

**12.**   Passing through the points $(5, 1)$ and $(2, -3)$

**13.**   $y$-intercept $4$ and $x$-intercept $-5$

**14.**   $x$-intercept $-2$ and $y$-intercept $-5$

**15.**   Parallel to the line $x + 4y - 2 = 0$ passing through $(4, 1)$

**16.**   Parallel to the line $2x + y = 0$ passing through $(7, 3)$

**17.**   Perpendicular to the line $4x + y - 2 = 0$ passing through the origin

**18.**   Perpendicular to the line $3x + 2y + 4 = 0$ passing through the point $(1, -3)$

**19.**   Passing through $(1, 3)$ and parallel to the line passing through $(4, 3)$ and $(-1, 2)$

**20.**   Passing through $(2, -3)$ and perpendicular to the line passing through $(-3, -5)$ and $(-1, -2)$

*Sketch the lines given in Problems 21–26.*

**21.**   $x + y + 4 = 0$       **22.**  $3x + y - 3 = 0$
**23.**   $2x + 3y - 1 = 0$     **24.**  $x = y$
**25.**   $x = 4$              **26.**  $y = -3$

**27.**   Find the $y$-intercept for $3x + 6y + 1 = 0$.
**28.**   Find the $x$-intercept for $4x + 3y - 2 = 0$.
**29.**   Find the slope and $y$-intercept for $Ax + By + C = 0$.

**30.**    Find the equation of the perpendicular bisector of the line segment connecting $(1, 4)$ and $(2, 5)$.

**B**    **31.**    If $A(0, 0)$, $B(6, 0)$, and $C(3, 4)$, show that $\triangle ABC$ is isosceles. Show that the median from $C$ is perpendicular to the base $AB$.

**32.**    Derive the formula for the distance between two points $P_1(x_1, y_1)$ and $P_2(x_2, y_2)$.

*For Problems 33–37, let $L_1$ and $L_2$ be nonvertical lines with slopes $m_1$ and $m_2$, respectively.*

**33.**    If $L_1 \parallel L_2$, prove $m_1 = m_2$.

**34.**    If $m_1 = m_2$, prove $L_1 \parallel L_2$.

**35.**    If $L_1 \perp L_2$, prove $m_1 m_2 = -1$.

**36.**    If $m_1 m_2 = -1$, prove $L_1 \perp L_2$.

**37.**    Show that the tangent of the angle $\theta$ measured in a positive direction from $L_1$ to $L_2$ is

$$\tan \theta = \frac{m_2 - m_1}{1 - m_1 m_2}$$

**38.**    If $P_1(x_1, y_1)$ and $P_2(x_2, y_2)$, show that the midpoint $M$ is given by

$$M = \left( \frac{x_1 + x_2}{2}, \frac{y_1 + y_2}{2} \right)$$

**39.**    Derive the equation of the line through $P_1(x_1, y_1)$ whose slope is $m$.

**40.**    Derive the equation of the line whose slope is $m$ and whose $y$-intercept is $(0, b)$.

**41.**    Prove that the lines joining the midpoints of the opposite sides of a quadrilateral bisect each other. *Hint:* Let $A(0, 0)$, $B(a, 0)$, $C(b, c)$, and $D(d, e)$ be the vertices of the quadrilateral.

**42.**    Prove that, if the diagonals of a parallelogram are equal, the parallelogram is a rectangle. *Hint:* Let $A(0, 0)$, $B(a, 0)$, $C(a + c, d)$, and $D(c, d)$ be the vertices of the parallelogram.

**43.**    Prove that the medians of a triangle meet at a point.

**44.**    Prove that the altitudes of a triangle meet at a point.

**45.**    Show that the distance between a point $(x_0, y_0)$ and a line $Ax + By + C = 0$ is

$$d = \frac{Ax_0 + By_0 + C}{\sqrt{A^2 + B^2}}$$

**46.**    Show that the midpoints of the sides of a quadrilateral are the vertices of a parallelogram.

# CHAPTER 2

# CONIC SECTIONS: PARABOLAS

## 2.1 Meeting a Conic

In Chapter 1 we considered lines characterized by equations that are *first* degree in two variables. In the next few chapters we will consider curves whose equations are second degree in two variables. It is possible to describe these curves as the intersection of a right circular cone and a plane (see Figure 10). The cone is thought of as extending infinitely in both directions. The part of the cone on each side of the vertex is called a *nappe,* and the side of the cone is called a *generator* of the cone.

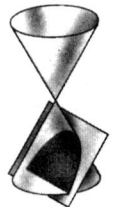

*Parabola*
The plane is parallel to one of the generators of the curve.

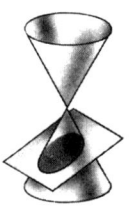

*Ellipse*
The plane intersects only one nappe. A circle is a special ellipse in which the plane is perpendicular to the axis of the cone.

*Hyperbola*
The plane intersects both nappes of the cone.

Figure 10  A conic section is the intersection of a plane and a cone. That is, the conics are the boundaries of the shaded regions.

These curves were studied extensively by the Greeks, but Descartes was the first to prove that every conic has an equation of the form

$$Ax^2 + Bxy + Cy^2 + Dx + Ey + F = 0$$

where $A$, $B$, $C$, $D$, $E$, and $F$ are constants. Notice if $A = B = C = 0$, the resulting equation describes a line.

## 2.2 The Parabola

**PARABOLA**

> A **parabola** is the set of all points in the plane equidistant from a given point (called the *focus*) and a given line (called the *directrix*).

In order to see what a parabola looks like, we use special graph paper like that shown in Figure 11, where $F$ is the focus and $L$ the directrix.

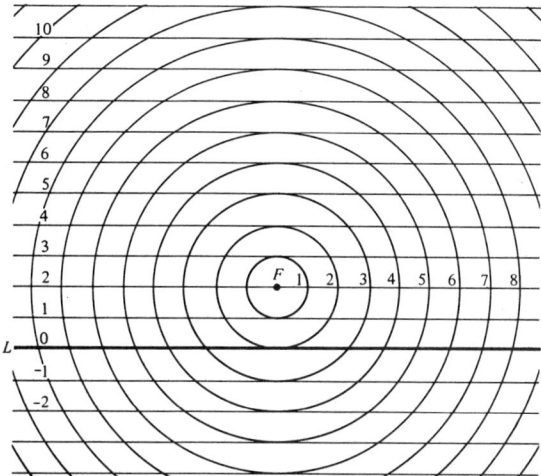

Figure 11 Parabola graph paper

To plot a parabola by using the definition, we need to plot all of the points in the plane that are equidistant from the focus and the directrix. Draw a line through the focus, perpendicular to the directrix. This line is called the **axis** of the parabola. Let $V$ be the point halfway between the focus and the directrix. This is the point of the parabola nearest to either the focus or the directrix. It is called the **vertex** or **center** of the parabola.<sup>*</sup> We plot other points equidistant from $F$ and $L$, as shown in Figure 12.

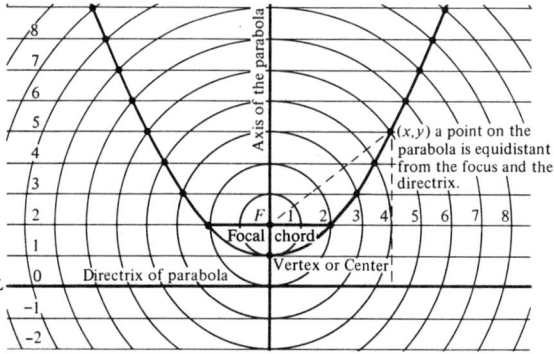

Figure 12

---

<sup>*</sup>You may be surprised that we use the word *center* to describe the vertex of a parabola. According to the *Mathematics Dictionary* by James and James, the word *center* is the point (if it exists) about which the curve is symmetrical. The vertex of a parabola satisfies this definition. Also, we will consistently use the notation $(h, k)$ to denote the center of all of the conic sections.

In Figure 12, let $c$ be the distance from the center to the focus. Notice that the distance from the center to the directrix is also $c$. If we consider the length of the segment that passes through the focus perpendicular to the axis and whose endpoints are on the parabola, we see that this segment has length $4c$. This segment is called the **focal chord.**

To obtain the equation of a parabola, we first consider a special case. Consider a parabola with focus $F(0, c)$ and directrix $y = -c$, where $c$ is any positive number. We have forced the parabola to have its center at the origin (remember that the center is halfway between the focus and the directrix). We have also forced the parabola to open up, as shown in Figure 13.

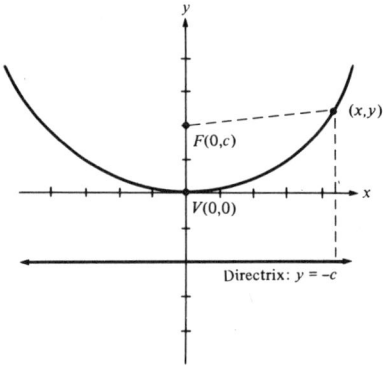

Figure 13  Graph of the parabola $x^2 = 4cy$

Let $(x, y)$ be any point on the parabola. Then, from the definition of a parabola, the distance from $(x, y)$ to $(0, c)$ is equal to the distance from $(x, y)$ to the directrix. Symbolically,

$$\sqrt{(x - 0)^2 + (y - c)^2} = |y + c|$$

Simplifing this equation, we obtain

$$x^2 = 4cy$$

which is the equation of this parabola.

We can repeat the argument above for parabolas that have their center at the origin and open down, left, or right, to obtain the results summarized in the following box. We are given a positive number $c$, which is the distance from the focus to the center.

**STANDARD-FORM PARABOLAS CENTERED AT (0, 0)**

| Parabola | Equation |
|---|---|
| opens up | $x^2 = 4cy$ |
| opens down | $x^2 = -4cy$ |
| opens right | $y^2 = 4cx$ |
| opens left | $y^2 = -4cx$ |

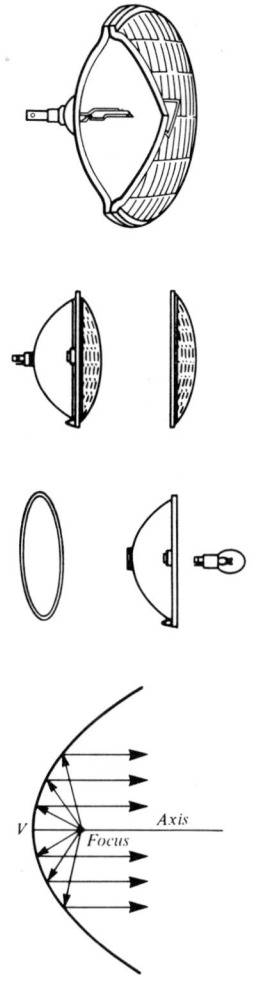

A parabolic reflector (or *parabolic mirror*) has the property that, if a source of light is placed at the *focus* of a mirror, the light rays will reflect from the mirror as rays parallel to the axis. Such a parabolic reflector is used in the automobile headlamp to deliver an intense concentrated beam of light.

For its source of light the headlamp has a filament to which electric wiring connections from the battery furnish a filament voltage.

For safer night driving, we do not want all light rays to be parallel to the axis. Some light must be aimed far down the road; some light must be spread to the side of the road; and some light must be distributed upward to illuminate high objects such as bridges or overhead signs.

How do we spread and aim light beams to these different directions? If we deliberately *offset* the filament from the focal point, we change the beam entirely. In our four-lamp system today, we have one-filament sealed beams in two headlamps, and two-filament sealed beam units in the other two. The position of the filaments in relation to the reflector accomplishes most of the desired illumination patterns. The rest of the light spreading is taken care of by the design of the lenses. These contain flutes and prisms designed to bend light rays.

From *Mathematics at Work in General Motors,* Number 3, courtesy of General Motors Corporation.

**EXAMPLE 1**

Graph $x^2 = 8y$.

*Solution*

We recognize this equation as a parabola that opens up. The center is (0, 0) and we notice by inspection that

$$4c = 8$$
$$c = 2$$

Thus, the focus is (0, 2). After we plot the center $V(0, 0)$ and the focus $F(0, 2)$, the only question is the "thickness" of the parabola. Remember that the *length of the focal chord is 4c*, so in this case it is 8. Since a parabola is symmetric with respect to its axis, we draw a segment of length 8 with the center at $F$. Now, using these three points (the center and the endpoints of the focal chord), we sketch the parabola, as shown in Figure 14.

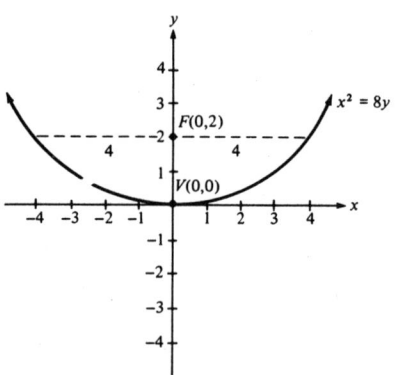

Figure 14  Graph of the parabola $x^2 = 8y$                □

**EXAMPLE 2**          Graph $y^2 = -12x$.

*Solution*          We recognize this equation as a parabola that opens left. The center is $(0, 0)$ and

$$4c = 12$$
$$c = 3$$

(recall that $c$ is positive), so the focus is $(-3, 0)$. The length of the focal chord is 12, and we draw the parabola as shown in Figure 15.

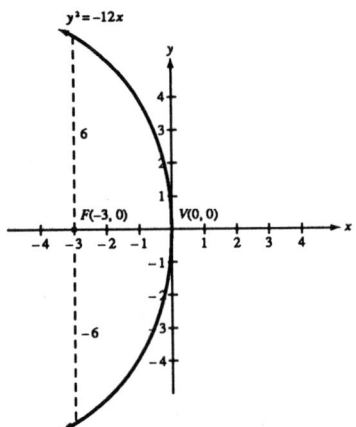

Figure 15  Graph of the parabola $y^2 = -12x$                □

**EXAMPLE 3**          Graph $2y^2 - 5x = 0$.

*Solution*          First we put the equation into standard algebraic form by solving for the second-degree term.

$$y^2 = \tfrac{5}{2} x$$

The center is $(0, 0)$, and

$$4c = \tfrac{5}{2}$$

Thus, the parabola opens to the right and the focus is $(\frac{5}{8}, 0)$, with the length of the focal chord $\frac{5}{2}$, as shown in Figure 16.

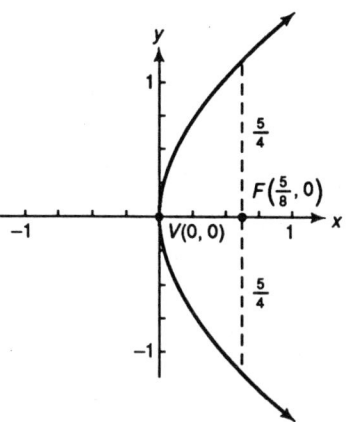

Figure 16  Graph of the parabola $2y^2 - 5x = 0$ ☐

**EXAMPLE 4**

Find the equation of the parabola with directrix $y = 4$ and focus at $F(0, -4)$. Sketch the graph.

*Solution*

This curve is a parabola that opens down, with center at the origin, as shown in Figure 17. We see that $c = 4$ and $4c = 16$. Since the equation is of the form

$$x^2 = -4cy$$

we see that the desired equation is

$$x^2 = -16y$$

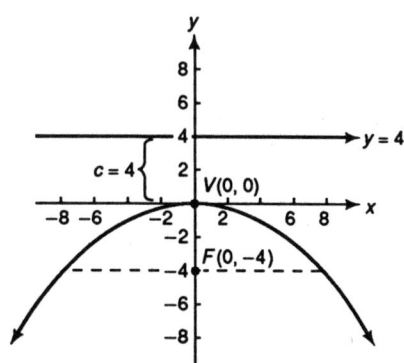

Figure 17  Graph of the parabola with focus at $(0, -4)$ and directrix $y = 4$ ☐

**EXAMPLE 5**

Find the equation of the parabola with directrix $x = 3$ and focus at $F(-3, 0)$. Graph the curve.

*Solution*     This curve is a parabola that opens to the left and is of the form

$$y^2 = -4cx$$

We see that $c = 3$, so the desired equation is

$$y^2 = -12x$$

and the graph is drawn as shown in Figure 18.

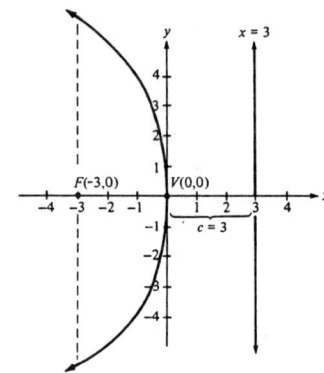

Figure 18  Graph of the parabola with
directrix $x = 3$ and focus at $(-3, 0)$          □

## 2.3  Translation of Axes

Thus far we have assumed that the center of the parabolas we have been considering is at the origin and the directrix is parallel to one of the coordinate axes. Suppose, however, that we are given a parabola with center at $(h, k)$ and a directrix parallel to one of the coordinate axes, as shown in Figure 19.

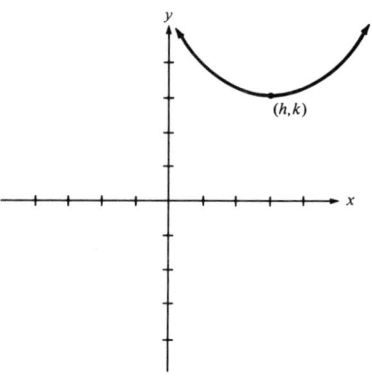

Figure 19

Even though we could find the equation of this parabola from the definition of a parabola, we will approach the problem from a different viewpoint — one that will help us not only with parabolas, but with the other conics as well.  Suppose we draw a new coordinate system with origin at $(h, k)$, as shown in Figure 20.

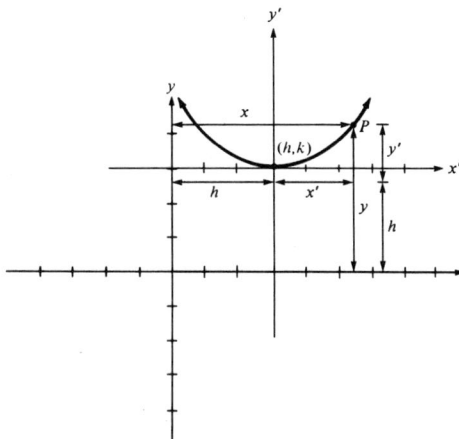

Figure 20    Translation of axes

Let $P$ be a point on the parabola.  The coordinates of $P$ are $(x, y)$ in the old coordinate system and $(x', y')$ in the new coordinate system.  The equation in the new coordinate system is seen to be

$$(x')^2 = 4cy'$$

To find the equation in the old coordinate system, we need to find the relationships between $x$, $x'$, $y$, and $y'$.  Notice that

$$x = x' + h \qquad \text{and} \qquad y = y' + k$$

Thus,

$$x' = x - h \qquad \text{and} \qquad y' = y - k$$

and we see (by substitution) that the equation in the old coordinate system is

$$(x - h)^2 = 4c(y - k)$$

Of course, this substitution applies regardless of the way the parabola opens, so we have the following general result, which you should learn.

**STANDARD-FORM PARABOLAS CENTERED AT $(h, k)$**

| Parabola | Equation |
|---|---|
| opens up | $(x - h)^2 = 4c(y - k)$ |
| opens down | $(x - h)^2 = -4c(y - k)$ |
| opens right | $(y - k)^2 = 4c(x - h)$ |
| opens left | $(y - k)^2 = -4c(x - h)$ |

**2.4  Problem 1:  Given the Equation, Graph the Curve**

**EXAMPLE 6**

Sketch $x^2 + 4y + 8x + 4 = 0$.

*Solution*

**Step 1:**  Bring together the terms involving the variable that has a squared term.

$$x^2 + 8x = 4y - 4$$

**Step 2:**  Complete the square for the variable that is squared.  See Appendix C for a review of the method of completing the square.

$$x^2 + 8x + \mathbf{16} = -4y - 4 + \mathbf{16}$$

**Step 3:**  Factor the side that is the perfect square, and also factor out the coefficient of the first-degree term.

$$(x + 4)^2 = -4(y - 3)$$

**Step 4:**  Determine the center (by inspection).  Plot $(h, k)$ as shown in Figure 21; in this example the center is $(-4, 3)$.

**Step 5:**  Determine the focus.  By inspection, $4c = 4$, $c = 1$, and the parabola opens down from the center, as shown in Figure 21.

**Step 6:**  Plot the endpoints of the focal chord; $4c = 4$.  We draw the parabola as shown in Figure 21.

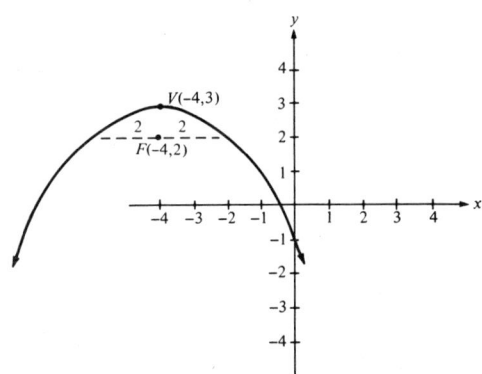

Figure 21  Graph of $x^2 + 4y + 8x + 4 = 0$            □

**EXAMPLE 7**

Sketch $2y^2 + 6y + 5x + 10 = 0$.

*Solution*

$$2y^2 + 6y = -5x - 10$$
$$y^2 + 3y = -\tfrac{5}{2}x - 5$$
$$y^2 + 3y + \tfrac{9}{4} = -\tfrac{5}{2}x - 5 + \tfrac{9}{4}$$
$$\left(y + \tfrac{3}{2}\right)^2 = -\tfrac{5}{2}x - \tfrac{11}{4}$$
$$\left(y + \tfrac{3}{2}\right)^2 = -\tfrac{5}{2}\left(x + \tfrac{11}{10}\right)$$

We see that the center is $\left(-\tfrac{11}{10}, -\tfrac{3}{2}\right)$; and $4c = \tfrac{5}{2}$, so $c = \tfrac{5}{8}$.  We sketch the curve as shown in Figure 22.

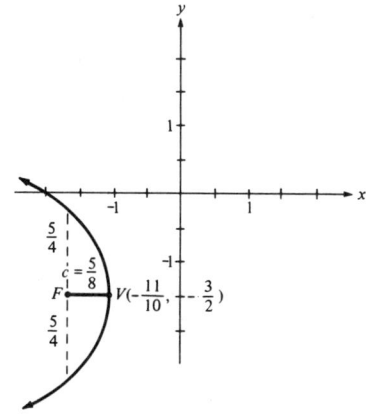

Figure 22   Graph of $2y^2 + 6y + 5x + 10 = 0$      □

## 2.5   Problem 2:   Given the Graph, Find the Equation

**EXAMPLE 8**

Find the equation of the parabola whose focus is at $(4, \; -3)$ and whose directrix is the line $x + 2 = 0$.

*Solution*

We sketch the given information as shown in Figure 23. The center is the point $(1, \; -3)$ since it must be equidistant from $F$ and the directrix. Also note that $c = 3$. Thus, we substitute into the equation

$$(y \; - \; k)^2 = 4c(x \; - \; h)$$

since the parabola opens to the right. The desired equation is

$$(y + 3)^2 = 12(x \; - \; 1)$$

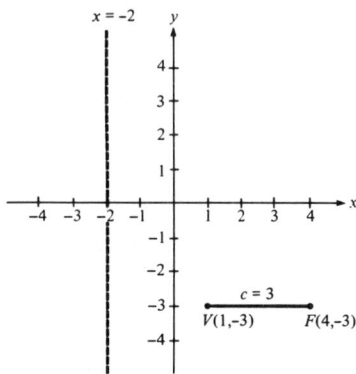

Figure 23   Sketch for Example 8      □

**EXAMPLE 9**

Find the equation of the parabola whose center is at $(2, -1)$, whose axis is parallel to the y-axis, and which passes through the point $(3, 2)$.

*Solution*

We sketch the given information as shown in Figure 24. We see that the parabola opens up and thus has the form

$$(x - h)^2 = 4c(y - k)$$

Since the center is $(2, -1)$, we have

$$(x - 2)^2 = 4c(y + 1)$$

Also, since it passes through $(3, 2)$, this point must satisfy the equation

$$(3 - 2)^2 = 4c(2 + 1)$$

Solving for $c$, we find

$$c = \tfrac{1}{12}$$

Therefore, the desired equation is

$$(x - 2)^2 = \tfrac{1}{3}(y + 1)$$

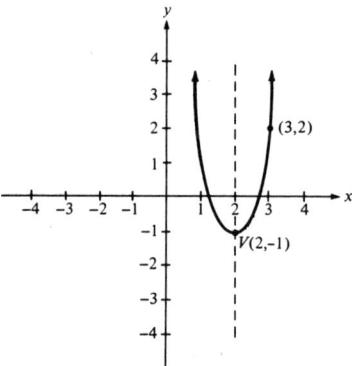

Figure 24   Sketch for Example 9                               □

---

### 2.6  PROBLEM SET 2

**A**

*Sketch the curves in Problems 1–22.*

1. $y^2 = 8x$
2. $y^2 = -12x$
3. $y^2 = -20x$
4. $4x^2 = 10y$
5. $3x^2 = -12y$
6. $2x^2 = -4y$
7. $2x^2 + 5y = 0$
8. $5y^2 + 15x = 0$
9. $3y^2 - 15x = 0$
10. $4y^2 + 3x = 12$
11. $(y - 1)^2 = 2(x + 2)$
12. $(y + 3)^2 = 3(x - 1)$

**13.** $(x + 2)^2 = 2(y - 1)$    **14.** $(x - 1)^2 = 3(y + 3)$

**15.** $(x - 1) = -2(y + 2)$    **16.** $(x + 3) = -3(y - 1)$

**17.** $y^2 - 4x + 10y + 13 = 0$    **18.** $y^2 + 4x - 3y + 1 = 0$

**19.** $2y^2 + 8y - 20x + 148 = 0$    **20.** $x^2 + 9y - 6x + 18 = 0$

**21.** $2y^2 + 8y - 3x + 15 = 0$    **22.** $3x^2 + 2x + 6y - 9 = 0$

*Find the equation of the curves in Problems 23–30. Sketch the curve.*

**23.**    Directrix $x = 0$ and focus at $(5, 0)$

**24.**    Directrix $y = 0$ and focus at $(0, -3)$

**25.**    Directrix $x - 3 = 0$ and vertex at $(-1, 2)$

**26.**    Directrix $y + 4 = 0$ and vertex at $(4, -1)$

**27.**    Vertex at $(-2, -3)$ and focus at $(-2, 3)$

**28.**    Vertex at $(-3, 4)$ and focus at $(1, 4)$

**29.**    The set of all points whose distances from $(4, 3)$ equal their distances from $(0, 3)$.

**30.**    The set of all points whose distances from $(4, 3)$ equal their distances from $(-2, 1)$.

**31.**    Find the equation of the parabola whose center is at $(-3, 2)$, whose axis is parallel to the $y$-axis, and which passes through $(-2, -1)$.

**32.**    Find the equation of the parabola whose center is at $(4, 2)$, whose axis is parallel to the $x$-axis, and which passes through $(-3, -4)$.

**33.**    Find the equation of the parabola with focus at $(4, -3)$ and directrix $x - y + 3 = 0$. *Hint:* Use the definition of a parabola and Problem 45 of Chapter 1.

**34.**    Find the equation of the parabola with focus at $(3, -5)$ and directrix $12x - 5y + 4 = 0$. *Hint:* Use the definition of a parabola and Problem 45 of Chapter 1.

**35.**    Suppose the path of a baseball that is hit follows a parabolic path that is 200 ft wide at the base and 50 ft high at the center. Write the equation of a parabola that gives the path of this baseball if we let the origin be the point of departure for the ball.

**B**

**36.**    Derive the equation of a parabola with $F(-c, 0)$, where $c$ is a positive number and the directrix is the line $y = c$.

**37.**    Derive the equation of a parabola with $F(0, c)$, where $c$ is a positive number and the directrix is the line $x = -c$.

**38.**    Derive the equation of a parabola with $F(0, -c)$ where $c$ is a positive number and the directrix is the line $x = c$.

**39.**    Show that the length of the focal chord for the parabola $y^2 = 4cx$ is $4c$.

# CHAPTER 3

# CONIC SECTIONS: ELLIPSES AND CIRCLES

### 3.1 The Ellipse

**ELLIPSE**

> An **ellipse** is the set of all points in the plane such that, for each point on the curve, the sum of its distances from two fixed points is a constant.

The fixed points are called the *foci* (plural of focus). To see what an ellipse looks like, we will use special graph paper like that shown in Figure 25, where $F_1$ and $F_2$ are the foci.

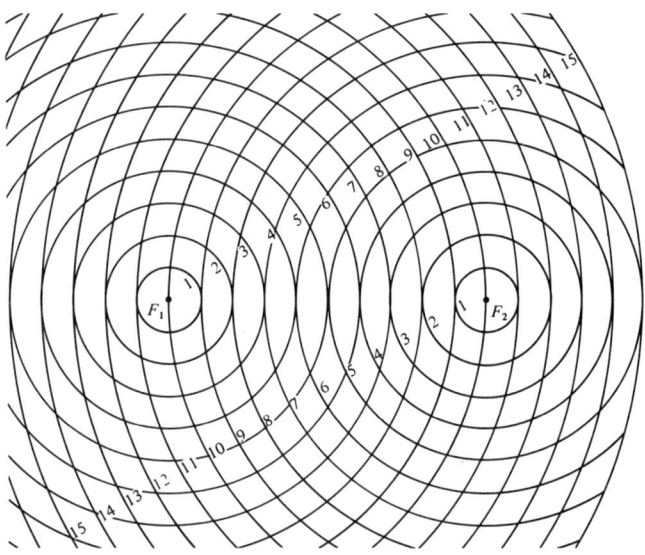

Figure 25 Ellipse graph paper

Let the given constant be 12. We then plot all of the points in the plane so that the sum of their distances from the foci is 12. For example, if a point is 8 units from $F_1$, it must be 4 units from $F_2$, and we plot the points $P_1$ and $P_2$ as shown in Figure 26.

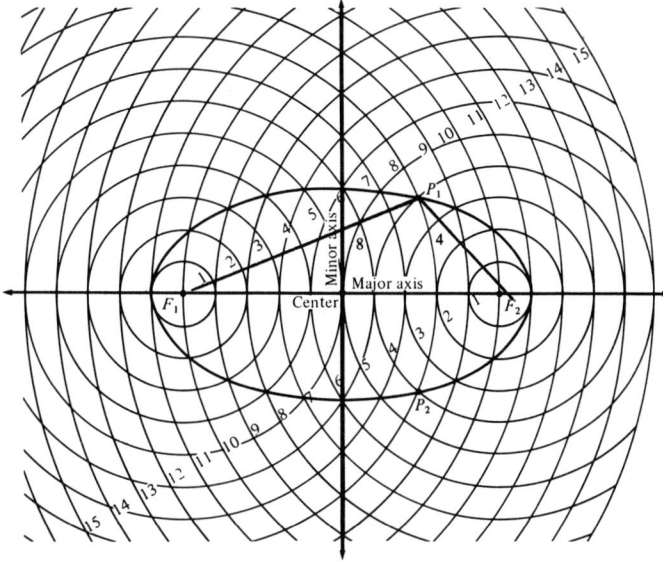

Figure 26

The line passing through $F_1$ and $F_2$ is called the **major axis**. The **center** is the midpoint of the segment connecting $F_1$ and $F_2$. The line passing through the center and perpendicular to the major axis is called the **minor axis**. We see that the ellipse is symmetric with respect to both the major and minor axes.

### 3.2 Focusing on the Equation

To find the equation of an ellipse, we first consider a special case. Let the distance from the center to a focus be the positive number $c$. That is, let $F_1(-c, 0)$ and $F_2(c, 0)$ be the focus, and let the constant distance be $2a$, as shown in Figure 27.

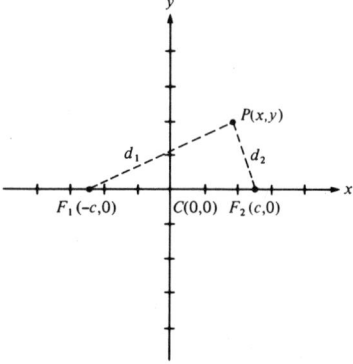

Figure 27

Notice that the center of this ellipse is $(0, 0)$. Let $P(x, y)$ be any point on the ellipse. By using the distance formula and the definition of an ellipse, we obtain

$$d_1 + d_2 = 2a$$

or

$$\sqrt{(x + c)^2 + (y - 0)^2} + \sqrt{(x - c)^2 + (y - 0)^2} = 2a$$

If we simplify this equation, we obtain

$$\frac{x^2}{a^2} + \frac{y^2}{b^2} = 1$$

If we let $x = 0$, we obtain the $y$-intercepts

$$0 + \frac{y^2}{b^2} = 1$$

$$y = \pm b$$

If we let $y = 0$, we obtain the $x$-intercepts $x = \pm a$. The intercepts on the major axis are called the *vertices* of the ellipse.

**EXAMPLE 1**        Sketch $\dfrac{x^2}{9} + \dfrac{y^2}{4} = 1$.

*Solution*        The center of the ellipse is $(0, 0)$, since this is the special case we considered above. The $x$-intercepts are at $\pm 3$ units from the center (these are the vertices), and the $y$-intercepts are at $\pm 2$ units from the center. We now sketch the ellipse as shown in Figure 28.

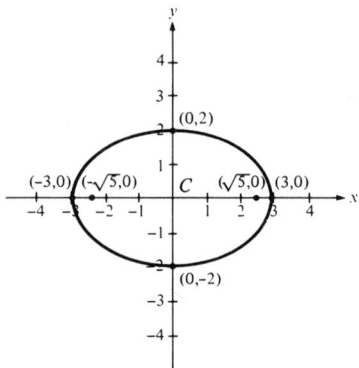

Figure 28   Graph of $\dfrac{x^2}{9} + \dfrac{y^2}{4} = 1$

The foci can be found, since

$$c^2 = a^2 - b^2$$

$$c^2 = 9 - 4$$

$$c = \pm\sqrt{5}$$

The equation of the ellipse whose major axis is vertical, with $F_1(0, c)$ and $F_2(0, -c)$ and constant distance $2a$, is found in similar fashion.

If we simplify the equation as before, we find

$$\frac{y^2}{a^2} + \frac{x^2}{b^2} = 1$$

and $b^2 = a^2 - c^2$. $\qquad\qquad\qquad\qquad\qquad\qquad\qquad$ □

Notice that in both cases $a^2$ must be larger than both $c^2$ and $b^2$. If it were not, we would have a square number equal to a negative number, which is not possible in the set of real numbers.

**EXAMPLE 2**

Sketch $\frac{x^2}{4} + \frac{y^2}{9} = 1$.

*Solution*

$a^2 = 9$ and $b^2 = 4$; this is an ellipse with a vertical major axis. The $x$-intercepts are $\pm 2$, and the $y$-intercepts are $\pm 3$ (these numbers locate the vertices). The sketch is shown in Figure 29.

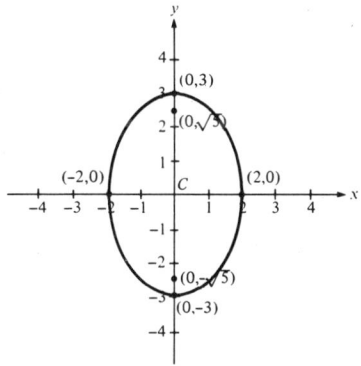

Figure 29  Graph of $\frac{x^2}{4} + \frac{y^2}{9} = 1$

The foci are found by

$$c^2 = a^2 - b^2$$
$$c^2 = 9 - 4$$
$$c = \pm\sqrt{5}$$

Remember, the foci are always on the major axis, so we plot $(0, \sqrt{5})$ and $(0, -\sqrt{5})$ for the foci. $\qquad\qquad\qquad\qquad\qquad\qquad$ □

If we let the center of the ellipse be the point $(h, k)$, we can translate the coordinate axes as we did for the parabola, where

$$x' = x - h \qquad \text{and} \qquad y' = y - k$$

The equations of the ellipses with center at $(h, k)$ are given below and should be learned.

| | Ellipse | Equation |
|---|---|---|
| **STANDARD-FORM ELLIPSES CENTERED AT $(h, k)$ WHERE $c^2 = a^2 - b^2$** | horizontal | $\dfrac{(x - h)^2}{a^2} + \dfrac{(y - k)^2}{b^2} = 1$ |
| | vertical | $\dfrac{(y - k)^2}{a^2} + \dfrac{(x - h)^2}{b^2} = 1$ |

### 3.3  Problem 1:  Given the Equation, Graph the Curve

**EXAMPLE 3**

Graph $\dfrac{(x - 3)^2}{25} + \dfrac{(y - 1)^2}{16} = 1$.

*Solution*

**Step 1:**  Plot the center $(h, k)$.  By inspection, the center of this ellipse is $(3, 1)$.   This becomes the center of a new translated coordinate system.  The vertices and foci are now measured with reference to the new origin at $(3, 1)$.

**Step 2:**  Plot the $x'$- and $y'$-intercepts.  These are located $\pm 5$ units and $\pm 4$ units from the center, respectively.   Remember to measure these distances from $(3, 1)$, as shown in Figure 30.

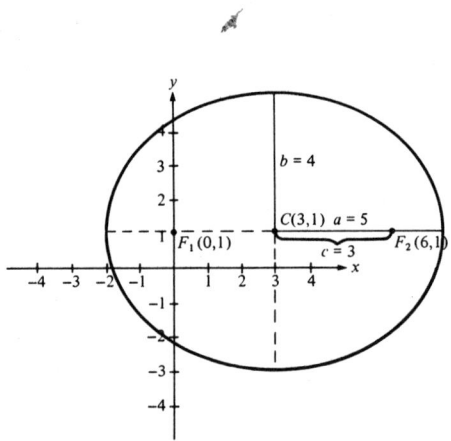

Figure 30  Graph of $\dfrac{(x - 3)^2}{25} + \dfrac{(y - 1)^2}{16} = 1$

The foci are found by

$$c^2 = a^2 - b^2$$
$$= 25 - 16$$
$$= 9$$

The distance from the center to the foci is 3, so the coordinates of the foci are $(6, 1)$ and $(0, 1)$.    □

The orbits of the planets are elliptical in shape. If the sun is placed at one of the foci of a giant ellipse, the orbit of the earth is an ellipse. The perihelion is the point where the planet comes closest to the sun and the aphelion is the farthest the planet is from the sun.

The eccentricity (see Section 2.4) of the planets measures the "roundness" of their orbits. The eccentricity of a circle is 0 and that of a parabola is 1. The eccentricity for the planets in our solar system are given at right.

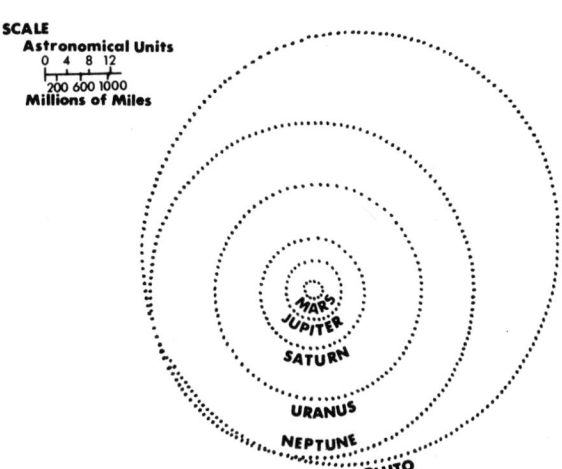

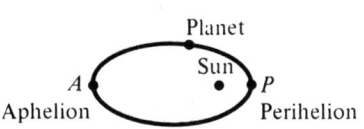

| Planet | Eccentricity |
|---|---|
| Mercury | .194 |
| Venus | .007 |
| Earth | .017 |
| Mars | .093 |
| Jupiter | .048 |
| Saturn | .056 |
| Uranus | .047 |
| Neptune | .009 |
| Pluto | .249 |

The orbit of a satellite can be calculated from Kepler's third law:

$$\frac{\text{mass of (planet + satellite)}}{\text{mass of (sun + planet + satellite)}} = \frac{(\text{semimajor axis of satellite orbit})^3}{(\text{semimajor axis of planet orbit})^3} \times \frac{(\text{period of planet})^2}{(\text{period of satellite})^2}$$

**EXAMPLE 4**

Graph $3x^2 + 4y^2 + 24x - 16y + 52 = 0$.

*Solution*

**Step 1:** Associate the $x$ and the $y$ terms:

$$(3x^2 + 24x) + (4y^2 - 16y) = -52$$

**Step 2:** We need to complete the square in *both* $x$ and $y$, and this requires that the coefficients of the squared terms be one. We can get the coefficients of the squared terms to be one by factoring:

$$3(x^2 + 8x \quad) + 4(y^2 - 4y \quad) = -52$$

Next, complete the square for *both* $x$ and $y$. Be sure to add the same number to both sides:

$$\text{Added 16 to both sides}$$

$$3(x^2 + 8x + 16) + 4(y^2 - 4y + 4) = -52 + 3 \cdot 16 + 4 \cdot 4$$

$$\text{------ Added 48 to both sides ------}$$

**Step 3:**  Factor:

$$3(x + 4)^2 + 4(y - 2)^2 = 12$$

**Step 4:**  Divide both sides by 12:

$$\frac{(x + 4)^2}{4} + \frac{(y - 2)^2}{3} = 1$$

**Step 5:**  Plot the center $(h, k)$.  By inspection, we see the center is $(-4, 2)$.  The vertices are $\pm 2$ units from the center and on the major axis.  The points of interception are $\pm \sqrt{3}$ units from the center, as shown in Figure 31.

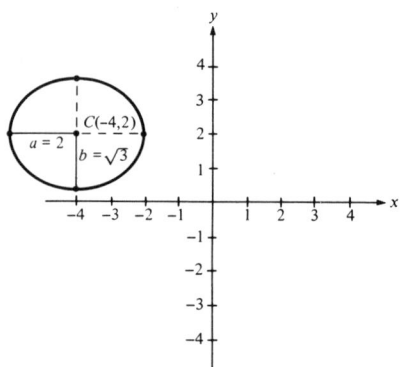

Figure 31  Graph of $3x^2 + 4y^2 + 24x - 16y + 52 = 0$     □

**EXAMPLE 5**

Graph $49x^2 + 9y^2 - 196x + 54y + 213 = 0$.

*Solution*

$$(49x^2 - 196x) + (9y^2 + 54y) = -213$$

$$49(x^2 - 4x \quad) + 9(y^2 + 6y \quad) = -213$$

$$49(x^2 - 4x + 4) + 9(y^2 + 6y + 9) = -213 + 4 \cdot 49 + 9 \cdot 9$$

$$49(x - 2)^2 + 9(y + 3)^2 = 64$$

$$\frac{49(x - 2)^2}{64} + \frac{9(y + 3)^2}{64} = 1$$

$$\frac{(x - 2)^2}{\frac{64}{49}} + \frac{(y + 3)^2}{\frac{64}{9}} = 1$$

We multiplied the first term by $\dfrac{\frac{1}{49}}{\frac{1}{49}}$ and the second term by $\dfrac{\frac{1}{9}}{\frac{1}{9}}$ for the last step.  We see that the center is $(2, -3)$, with vertices a distance of $\pm \frac{8}{3}$ on the major axis; on the minor axis the curve passes through points $\pm \frac{8}{7}$ from the center, as shown in Figure 32.

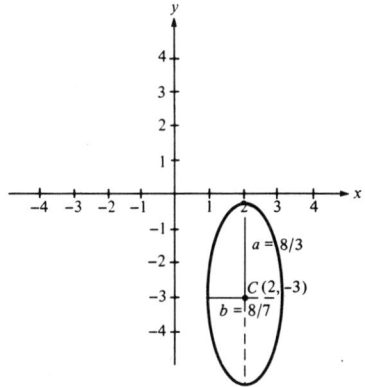

**Figure 32** Graph of $49x^2 + 9y^2 - 196x + 54y + 213 = 0$ □

### 3.4 Circles

We have seen that some ellipses are more circular than others. An ellipse's "flatness" is called its *eccentricity* and is defined as

$$\epsilon = \frac{c}{a}$$

Notice that

$$\epsilon = \frac{c}{a} = \frac{\sqrt{a^2 - b^2}}{a} = \sqrt{\frac{a^2 - b^2}{a^2}} = \sqrt{1 - \left(\frac{b}{a}\right)^2}$$

Since $c < a$, we see that $\epsilon$ is between 0 and 1. If $a = b$, then $\epsilon = 0$ and the conic is a **circle**. If the ration $\frac{a}{b}$ is small, then the ellipse is very flat. Thus, for an ellipse,

$$0 \le \epsilon < 1$$

and $\epsilon$ measures the amount of "roundness" of the ellipse. The closer $\epsilon$ is to 1, the flatter the ellipse becomes.

Let us take a closer look at the case where $a = b$. This common distance is called the *radius* of the circle and is denoted $r$. Thus, when $a = b = r$, we have

$$\frac{(x - h)^2}{r^2} + \frac{(y - k)^2}{r^2} = 1$$

If we multiply both sides by $r^2$, we obtain what is called the standard form of a circle with center $(h, k)$ and radius $r$.

**STANDARD-FORM CIRCLE CENTERED AT $(h, k)$**

$$(x - h)^2 + (y - k)^2 = r^2$$

**EXAMPLE 6**

*Solution*

Graph $x^2 + y^2 + 6x - 14y + 22 = 0$.

We complete the square in $x$ and $y$:

$$(x^2 + 6x \quad) + (y^2 - 14y \quad) = -22$$
$$(x^2 + 6x + 9) + (y^2 - 14y + 49) = -22 + 9 + 49$$
$$(x + 3)^2 + (y - 7)^2 = 36$$

We see that this is a circle with center at $(-3, 7)$ and radius 6, as shown in Figure 33.

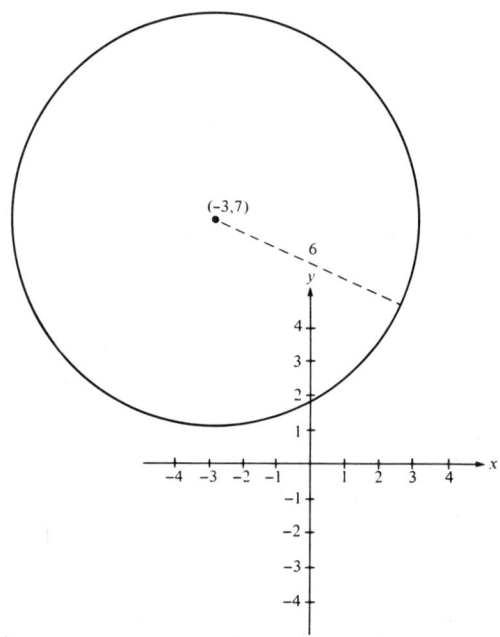

Figure 33  Graph of $x^2 + y^2 + 6x - 14y + 22 = 0$     □

**EXAMPLE 7**

*Solution*

Find the equation of the circle passing through the points $(-3, 4)$, $(4, 5)$, and $(1, -4)$. Sketch the graph.

Now $(x - h)^2 + (y - k)^2 = r^2$ can be written as

$$x^2 + y^2 + C_1 x + C_2 y + C_3 = 0$$

for some constants $C_1$, $C_2$, and $C_3$. Since the points lie on the circle, they satisfy the equation. Thus,

$$(-3, 4): \qquad -3C_1 + 4C_2 + C_3 = -(-3)^2 - (4)^2 = -25$$
$$(4,5): \qquad 4C_1 + 5C_2 + C_3 = -(4)^2 - (5)^2 = -41$$
$$(1, -4): \qquad C_1 - 4C_2 + C_3 = -(1)^2 - (-4)^2 = -17$$

We have a system of three equations and three unknowns that we solve simultaneously to find $C_1 = -2$, $C = -2$, and $C_3 = -23$. Thus (see Figure 34),

$$x^2 + y^2 - 2x - 2y - 23 = 0$$
$$(x^2 - 2x + 1) + (y^2 - 2y + 1) = 23 + 1 + 1$$
$$(x - 1)^2 + (y - 1)^2 = 25$$

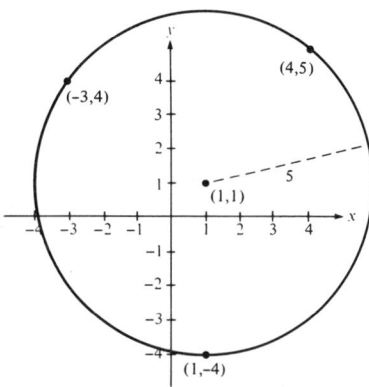

Figure 34  Sketch for Example 7 ☐

### 3.5  Problem 2:  Given the Graph, Find the Equation

**EXAMPLE 8**

Find the equation of the ellipse with vertices at ($\pm 5$, 0) and foci at ($\pm 3$, 0).

*Solution*

By inspection, the ellipse is horizontal and is centered at the origin, where $a = 5$ and $c = 3$.  Thus,

$$c^2 = a^2 - b^2$$
$$9 = 25 - b^2$$
$$b^2 = 16$$

The equation is
$$\frac{x^2}{25} + \frac{y^2}{16} = 1 \qquad\qquad ☐$$

**EXAMPLE 9**

Find the equation of the ellipse with vertices at (3, 2) and (3, $-4$) and foci at (3, $\sqrt{5} - 1$) and (3, $-\sqrt{5} - 1$).

*Solution*

The relationship among the variables is shown in Figure 35.  We see that $a = 3$ and $c = \sqrt{5}$.  Thus,

$$c^2 = a^2 - b^2$$
$$5 = 9 - b^2$$
$$b^2 = 4$$

The equation is

$$\frac{(y+1)^2}{9} + \frac{(x-3)^2}{4} = 1$$

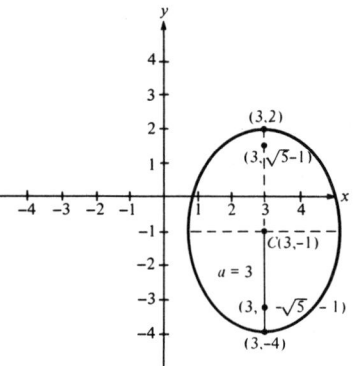

Figure 35  Sketch for Example 9    □

**EXAMPLE 10**

Find the equation that represents the set of all points the sum of whose distances from $(-3, 2)$ and $(5, 2)$ is 16.

*Solution*

By inspection, the ellipse is horizontal and is centered at $(1, 2)$. We are given

$$2a = 16$$

$$a = 8$$

and we see that $c = 4$. Thus,

$$c^2 = a^2 - b^2$$

$$16 = 64 - b^2$$

$$b^2 = 48$$

The equation is

$$\frac{(x-1)^2}{64} + \frac{(y-2)^2}{48} = 1$$    □

**EXAMPLE 11**

Find the equation of the curve with foci at $(-3, 6)$ and $(-3, 2)$, with eccentricity $\frac{1}{5}$.

*Solution*

By inspection, the ellipse is vertical and is centered at $(-3, 4)$ with $c = 2$. Since

$$\epsilon = \frac{c}{a} = \frac{1}{5}$$

and $c = 2$, we have

$$\frac{2}{a} = \frac{1}{5}$$

which implies $a = 10$. *   Also,

---

*Note that it is NOT correct to say that $\frac{c}{a} = \frac{1}{5}$ implies $c = 1$ and $a = 5$.

$$c^2 = a^2 - b^2$$
$$4 = 100 - b^2$$
$$b^2 = 96$$

Thus, the equation is

$$\frac{(y-4)^2}{100} + \frac{(x+3)^2}{96} = 1 \qquad\qquad \square$$

## 3.6 PROBLEM SET 3

**A**

*Sketch the curves in Problems 1–18.*

1. $x^2 + y^2 = 1$
2. $25x^2 + 16y^2 = 400$
3. $\frac{x^2}{4} + \frac{y^2}{9} = 1$
4. $\frac{x^2}{25} + \frac{y^2}{36} = 1$
5. $36x^2 + 25y^2 = 900$
6. $3x^2 + 2y^2 = 6$
7. $4x^2 + 3y^2 = 12$
8. $5x^2 + 10y^2 = 7$
9. $(x-2)^2 + (y+3)^2 = 25$
10. $3(x+1)^2 + 4(y-1)^2 = 12$
11. $\frac{(x+3)^2}{81} + \frac{(y-1)^2}{9} = 1$
12. $\frac{(x-3)^2}{16} + \frac{(y-2)^2}{9} = 1$
13. $x^2 + 4x + y^2 + 6y - 12 = 0$
14. $9x^2 + 4y^2 - 18x + 16y - 11 = 0$
15. $16x^2 + 9y^2 + 96x - 36y + 36 = 0$
16. $x^2 + 25y^2 + 14x + 150y + 273 = 0$
17. $3x^2 + 4y^2 + 2x - 8y + 4 = 0$
18. $144x^2 + 72y^2 - 72x + 48y - 7 = 0$

*Find the equations of the curves in Problems 19–26. Sketch the curves.*

19. The set of points 6 units from the point $(4, 5)$.
20. The set of points such that the sum of distances from $(-6, 0)$ and $(6, 0)$ is 20.
21. The ellipse with vertices at $(0, 7)$ and $(0, -7)$ and foci at $(0, 5)$ and $(0, -5)$.
22. The ellipse with vertices at $(4, 3)$ and $(4, -5)$ and foci at $(4, 2)$ and $(4, -4)$.
23. The circle with center at $(-3, 2)$ tangent to the line $4x - 3y - 2 = 0$. See Problem 45, Chapter 1.
24. The ellipse with foci at $(-4, -3)$ and $(2, -3)$, with eccentricity $\frac{4}{5}$.
25. The circle passing through $(2, 2)$, $(-2, -6)$, and $(5, 1)$.
26. The ellipse passing through $(5, 2)$ and $(3, \sqrt{5})$, with axes along the coordinate axes.

**B**    **27.**    Derive the equation of the ellipse with foci at $(-c, 0)$ and $(c, 0)$ and constant distance $2a$. Let $b^2 = a^2 - c^2$. Show all of your work.

**28.**    Derive the equation of the ellipse with foci at $(0, c)$ and $(0, -c)$ and constant distance $2a$. Let $b^2 = a^2 - c^2$. Show all of your work.

**29.**    Derive the equation of a circle with center $(h, k)$ and radius $r$ by using the distance formula.

**30.**    If we are given an ellipse with foci at $(-c, 0)$ and $(c, 0)$ and vertices at $(-a, 0)$ and $(a, 0)$, we define the *directrices* of the ellipse as the lines $x = \frac{a}{\epsilon}$ and $x = \frac{-a}{\epsilon}$. Show that an ellipse is the set of all points whose distances from $F(c, 0)$ are equal to $\epsilon$ times their distances from the line $x = \frac{a}{\epsilon}$.

**31.**    A line segment through a focus parallel to a directrix and cut off by the ellipse is called the *focal chord*. Show that the length of the focal chord of

$$\frac{x^2}{a^2} + \frac{y^2}{b^2} = 1$$

is $\frac{2b^2}{a}$.

# CHAPTER 4

# CONIC SECTIONS: HYPERBOLAS

## 4.1 The Hyperbola

**HYPERBOLA**

A **hyperbola** is the set of all points in the plane such that, for each point on the curve, the difference of its distances from two fixed points is a constant.

The fixed points are called the *foci*. A hyperbola with foci at $F_1$ and $F_2$, where the given constant is 8, is shown in Figure 36. The line passing through the foci is called the *transverse axis*. The *center* is the midpoint of the segment connecting the foci. The line passing through the center perpendicular to the transverse axis is called the *conjugate axis*. We see that the hyperbola is symmetric with respect to both the transverse and conjugate axes.

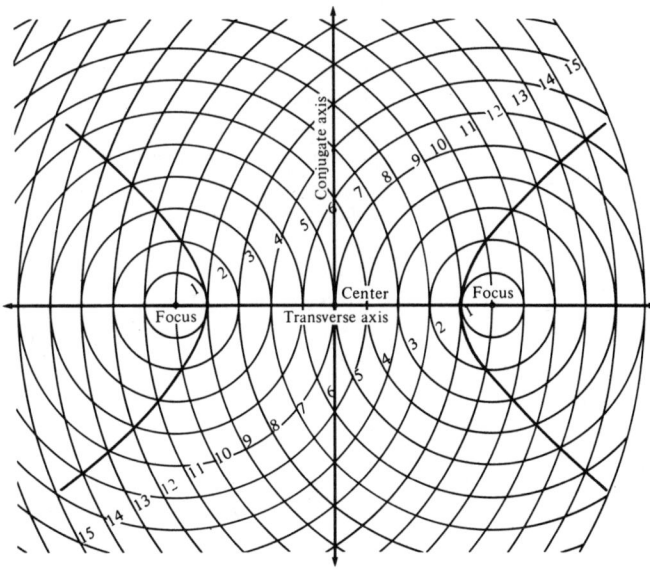

Figure 36 Hyperbolas plotted on hyperbola graph paper

If we use the definition, we can derive the equation of a hyperbola.

**STANDARD-FORM HYPERBOLA WITH CENTER AT (0, 0) WHERE $b^2 = c^2 - a^2$**

| Hyperbola | Equation |
|-----------|----------|
| horizontal | $\dfrac{x^2}{a^2} + \dfrac{y^2}{b^2} = 1$ |
| vertical | $\dfrac{y^2}{a^2} + \dfrac{x^2}{b^2} = 1$ |

Notice that for the ellipse $c^2 = a^2 - b^2$ and for the hyperbola $c^2 = a^2 - b^2$. When comparing $a$ and $c$, you should note that for the ellipse $a$ is the larger number, but for the hyperbola $c$ is the larger number. We see that the relative sizes of $a$ and $b$ are not relevant for a hyperbola, whereas for the ellipse $a^2 > b^2$. Also, the *eccentricity* for a hyperbola is $\epsilon = \frac{c}{a}$. Since $c > a$, we see that for a hyperbola $\epsilon > 1$.

### 4.2  Asymptotes

As with the other conics, we sketch the hyperbola by determining some information about the curve directly from the equation, by inspection. The points of intersection of the hyperbola with the transverse axis are called the *vertices*. For the curves with the equations

$$\frac{x^2}{a^2} - \frac{y^2}{b^2} = 1 \quad \text{and} \quad \frac{y^2}{a^2} - \frac{x^2}{b^2} = 1$$

we see that vertices occur at $(a, 0)$, $(-a, 0)$ and $(0, a)$, $(0, -a)$, respectively. The number $2a$ is the *length of the transverse axis*. The hyperbola does not intersect the conjugate axis, but if we plot the points $(0, b)$, $(0, -b)$, $(b, 0)$, and $(-b, 0)$, respectively, we determine a segment on the conjugate axis called the *length of the conjugate axis*.

**EXAMPLE 1**

Sketch $\dfrac{x^2}{4} - \dfrac{y^2}{9} = 1$.

*Solution*

The center of the hyperbola is $(0, 0)$, and $a = 2$ and $b = 3$. We plot the vertices $\pm 2$ units from the center, as shown in Figure 37. The transverse axis is along the $x$-axis, and the conjugate axis is along the $y$-axis.

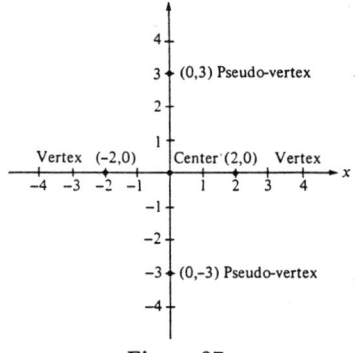

Figure 37

We plot the length of the conjugate axis $\pm 3$ units from the center. We call these points the **pseudo-vertices,** since the curve does not actually pass through these points. Next, we draw lines through the vertices and pseudo-vertices parallel to the axes of the hyperbola. These lines form what we will call the *central rectangle.* The lines passing through the corners of the central rectangle become our *guiding lines* for sketching the hyperbola, as shown in Figure 38.

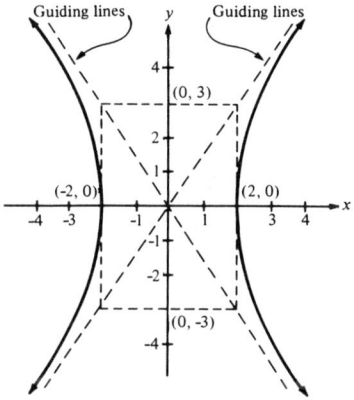

Figure 38   Graph of $\dfrac{x^2}{4} - \dfrac{y^2}{9} = 1$

We can find the equations of the guiding lines for a hyperbola

$$\frac{x^2}{a^2} - \frac{y^2}{b^2} = 1$$

to be

$$y = \frac{b}{a}x \quad \text{and} \quad y = -\frac{b}{a}x$$

It can be shown that, as a point $P$ moves away from the vertex of the hyperbola, the distance between the guiding lines and the hyperbola decreases. A line with this property is called an **asymptote** to the curve. Thus, the guiding lines we drew above are called the *asymptotes of the hyperbola.*

If we let the center of the hyperbola be the point $(h, k)$, we find the following equations for a hyperbola.

---

**STANDARD-FORM HYPERBOLA WITH CENTER AT $(h, k)$ WHERE $b^2 = c^2 - a^2$**

| Hyperbola | Equation |
|---|---|
| horizontal | $\dfrac{(x - h)^2}{a^2} + \dfrac{(y - k)^2}{b^2} = 1$ |
| vertical | $\dfrac{(y - k)^2}{a^2} + \dfrac{(x - h)^2}{b^2} = 1$ |

Hyperbolas are used in some techniques of navigation and of satellite tracking, where the difference of distances from two fixed points is observed (a ship's distances from two fixed stations on the shore, a satellite's distances from two antennas). If $A$ and $B$ are land stations and the difference in distance from them is electronically observed on the ship, its location must be on a certain hyperbola, with $A$ and $B$ as foci. If this procedure is then repeated for station $B$ along with a third station $C$, the ship must be on a hyperbola with these points as foci. Now, the *two sets of hyperbolas serve as a coordinate system* to locate the ship in the same way that it could be located on a Cartesian coordinate system.

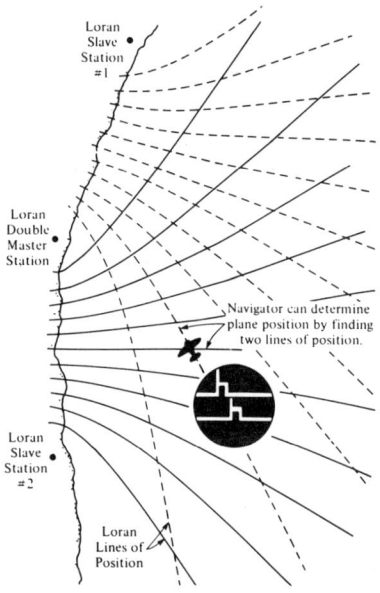

Loran measures the difference in the time of arrival of signals from two sets of stations. The plane's position at the intersection lines is charted on a special map based on a hyperbolic coordinate system. (Diagram courtesy of Bendix Corporation.)

### 4.3  Problem 1:  Given the Equation, Graph the Curve

**EXAMPLE 2**

Sketch $16x^2 - 9y^2 - 128x - 18y + 103 = 0$

*Solution*

Associate the $x$ and $y$ terms separately, and then complete the square.

$$(16x^2 - 128x) + (-9y^2 - 18y) = -103$$

$$16(x^2 - 8x\quad) - 9(y^2 + 2y\quad) = -103$$

$$16(x^2 - 8x + \mathbf{16}) - 9(y^2 + 2y + \mathbf{1}) = -103 + \mathbf{16 \cdot 16} - \mathbf{9 \cdot 1}$$

$$16(x - 4)^2 - 9(y + 1)^2 = 144$$

$$\frac{(x - 4)^2}{9} - \frac{(y + 1)^2}{16} = 1$$

The graph is shown in Figure 39.

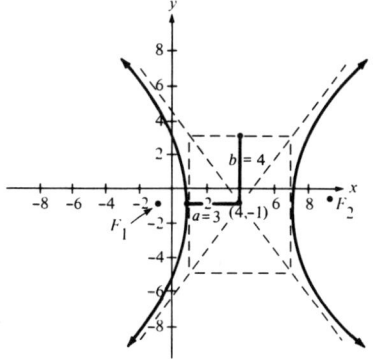

Figure 39  Sketch of $16x^2 - 9y^2 - 128x - 18y + 103 = 0$ ☐

We can also find the foci in Example 2, since
$$c^2 = a^2 + b^2$$
Thus,
$$c^2 = 9 + 16$$
$$c = \pm 5$$

### 4.4  Problem 2:  Given the Graph, Find the Equation

**EXAMPLE 3**

Find the equation of the hyperbola with vertices at $(2, 4)$ and $(2, -2)$ and foci at $(2, 6)$ and $(2, -4)$.

*Solution*

We plot the given points as shown in Figure 40.  We see that the center of the hyperbola is $(2, 1)$, since it is the midpoint of the segment connecting the foci.  We also see that $c = 5$ and $a = 3$. Thus, since
$$c^2 = a^2 + b^2$$
we have
$$25 = 9 + b^2$$
$$b^2 = 16$$
and the equation is
$$\frac{(y - 1)^2}{9} - \frac{(x - 2)^2}{16} = 1$$

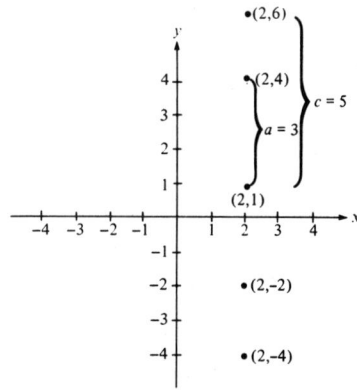

Figure 40 ☐

**EXAMPLE 4**

Find the equation of the hyperbola with foci at $(-3, 2)$ and $(5, 2)$ and with eccentricity $\frac{3}{2}$.

*Solution*

The center of the hyperbola is $(1, 2)$, and $c = 4$. Also, since

$$\epsilon = \frac{c}{a} = \frac{3}{2}$$

we have

$$\frac{4}{a} = \frac{3}{2}$$

which implies

$$a = \frac{8}{3}$$

Since $c^2 = a^2 + b^2$,

$$16 = \frac{64}{9} + b^2$$

$$b^2 = \frac{80}{9}$$

Thus, the equation is

$$\frac{(x - 1)^2}{\frac{64}{9}} - \frac{(y - 2)^2}{\frac{80}{9}} = 1 \qquad \square$$

**EXAMPLE 5**

Find the set of points such that the difference of their distances from $(6, 2)$ and $(6, -5)$ is always 3.

*Solution*

The center of the hyperbola is $(6, -\frac{3}{2})$, and $c = \frac{7}{2}$.

Also, $2a = 3$, so $a = \frac{3}{2}$.

Thus, since

$$c^2 = a^2 + b^2$$

$$\frac{49}{4} = \frac{9}{4} + b^2$$

$$b^2 = 10$$

The equation is

$$\frac{(y + \frac{3}{2})^2}{\frac{9}{4}} - \frac{(x - 6)^2}{10} = 1 \qquad \square$$

## 4.5  PROBLEM SET 4

**A**

*Sketch the curves in Problems 1–18.*

1. $x^2 - y^2 = 1$

2. $x^2 - y^2 = 4$

3. $\frac{x^2}{9} - \frac{y^2}{4} = 1$

4. $\frac{x^2}{4} - \frac{y^2}{9} = 1$

5. $\frac{y^2}{9} - \frac{x^2}{4} = 1$

6. $\frac{y^2}{4} - \frac{x^2}{9} = 1$

7. $36y^2 - 25x^2 = 900$

8. $3x^2 - 4y^2 = 5$

9. $\frac{(x - 2)^2}{4} - \frac{(y + 3)^2}{16} = 1$

10. $\frac{(x + 3)^2}{8} - \frac{(y - 1)^2}{5} = 1$

11. $\frac{(y + 2)^2}{6} - \frac{(x + 2)^2}{8} = 1$

12. $\frac{(y + 2)^2}{25} - \frac{(x + 1)^2}{16} = 1$

13. $5(x - 2)^2 - 2(y + 3)^2 = 10$

14. $4(x + 4)^2 - 3(y + 3)^2 = -12$

15. $3x^2 - 5y^2 + 18x + 10y - 8 = 0$

16. $9x^2 - 18x - 11 = 4y^2 + 16y$

17. $4y^2 - 8y + 4 = 3x^2 - 2x$

18. $4x^2 - 3y^2 - 24 - 112 = 0$

*Find the equation of the curves in Problems 19–24.*

19. The hyperbola with vertices at $(0, 5)$ and $(0, -5)$ and foci at $(0, 7)$ and $(0, -7)$.

20. The set of points such that the difference of their distances from $(-6, 0)$ and $(6, 0)$ is 10.

21. The hyperbola with foci at $(5, 0)$ and $(-5, 0)$ and with eccentricity 5.

22. The hyperbola with vertices at $(4, 4)$ and $(4, 8)$ and foci at $(4, 3)$ and $(4, 9)$.

23. The set of points such that the difference of their distances from $(4, -3)$ and $(-4, -3)$ is 6.

24. The hyperbola with vertices at $(-2, 0)$ and $(6, 0)$, passing through $(10, 3)$.

**B** 25. Derive the equation of the hyperbola with foci at $(-c, 0)$ and $(c, 0)$ and constant distance $2a$. Let $b^2 = c^2 - a^2$. Show all of your work.

26. Derive the equation of the hyperbola with foci at $(0, c)$ and $(0, -c)$ and constant distance $2a$. Let $b^2 = c^2 - a^2$. Show all of your work.

27. Given the hyperbola

$$\frac{x^2}{a^2} - \frac{y^2}{b^2} = 1$$

we define the *directrices* of the hyperbola as the lines $x = \frac{a}{\epsilon}$ and $x = -\frac{a}{\epsilon}$. Show that the hyperbola is the set of all points whose distances from $F(c, 0)$ are equal to $\epsilon$ times their distances from the line $x = \frac{a}{\epsilon}$.

28. Show that the asymptotes for the hyperbola

$$\frac{x^2}{a^2} - \frac{y^2}{b^2} = 1$$

have the equations $y = \frac{b}{a}x$ and $y = -\frac{b}{a}x$.

29. Show that the length of the diagonal of the central rectangle of a hyperbola is $2c$.

30. A line through a focus parallel to a directrix and cut off by the hyperbola is called the *focal chord*. Show that the length of the focal chord of

$$\frac{x^2}{a^2} - \frac{y^2}{b^2} = 1$$

is $\frac{2b^2}{a}$, and compare this to the result for an ellipse (Problem 31 of Chapter 3).

# CHAPTER 5

# CONIC SECTIONS: ROTATIONS

## 5.1 Degenerate Conics

Certain conics are classified as **degenerate.** To visualize some of these degenerate conics, reconsider Figure 10 which is repeated here for easy reference.

  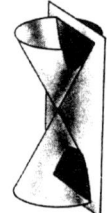

*Parabola*
The plane is parallel to one of the generators of the curve.

*Ellipse*
The plane intersects only one nappe. A circle is a special ellipse in which the plane is perpendicular to the axis of the cone.

*Hyperbola*
The plane intersects both nappes of the cone.

Figure 41  Conic sections

For a **degenerate parabola,** visualize the cone (Figure 41a) situated so that one of its generators lies in the plane; a line results. For a **degenerate ellipse,** visualize the plane intersecting at the vertex of the upper and lower nappes (see Figure 41b); a point results. And finally, for a **degenerate hyperbola,** visualize the plane situated so that the axis of the cone lies in the plane (see Figure 41c); a pair of intersecting lines results.

**EXAMPLE 1**

Sketch $\dfrac{(x - 2)^2}{4} + \dfrac{(y + 3)^2}{9} = 0$.

*Solution*

The only point that satisfies this equation is $(2, -3)$. This is an example of a *degenerate ellipse.* Notice that, except for the zero, the equation has the "form of an ellipse," as shown in Figure 42.

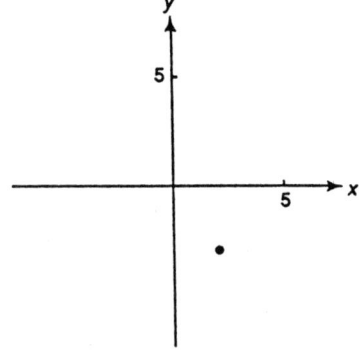

Figure 42  Graph of a degenerate ellipse

**EXAMPLE 2**

Sketch $\dfrac{x^2}{4} - \dfrac{y^2}{9} = 0$.

*Solution*

This equation has the "form of a hyperbola," but because of the zero it cannot be put into standard form. You can, however, treat this in factored form.

$$\frac{x^2}{4} - \frac{y^2}{9} = 0$$

$$\left(\frac{x}{2} - \frac{y}{3}\right)\left(\frac{x}{2} + \frac{y}{3}\right) = 0$$

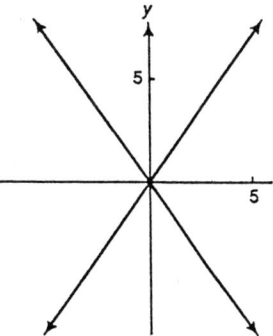

**Figure 43** Graph of a degenerate hyperbola

This is true when either factor is zero:

$$\frac{x}{2} - \frac{y}{3} = 0 \quad \text{or} \quad \frac{x}{2} + \frac{y}{3} = 0$$

The graph (singular) consists of two lines and is called a *degenerate hyperbola*.    □

### 5.2 The General Second-degree Equation

All of the curves we have considered can be characterized by the general second-degree equation

$$Ax^2 + Bxy + Cy^2 + Dx + Ey + F = 0$$

Notice that the $xy$ term has not appeared before in this book. The presence of this term indicates that the conic has been rotated.

It is important to be able to recognize the curves by inspection of the equation before we begin. The first thing to notice is whether or not there is an $xy$-term.

**If $B = 0$ (that is, there is no $xy$-term):**

| | Type of Curve | Degree of Equation | Degree in $x$ | Degree in $y$ | Relationship to General Equation |
|---|---|---|---|---|---|
| 1. | line | first | first | first | $A = C = 0$ |
| 2. | parabola | second | first | second | $A = 0$ and $C \neq 0$ |
| | | | second | first | $A \neq 0$ and $C = 0$ |
| 3. | ellipse | second | second | second | $A$ and $C$ have the same sign |
| 4. | circle | second | second | second | $A = C$ |
| 5. | hyperbola | second | second | second | $A$ and $C$ have opposite signs |

**If $B \neq 0$ (that is, there is an $xy$-term):**

| | Discriminant | Type of Curve |
|---|---|---|
| 1. | $B^2 - 4AC < 0$ | ellipse |
| 2. | $B^2 - 4AC = 0$ | parabola |
| 3. | $B^2 - 4AC > 0$ | hyperbola |

**EXAMPLE 3**

Identify each of the given curves.

**a.** $x^2 + 4xy + 4y^2 = 9$

**b.** $2x^2 + 3xy + y^2 = 25$

**c.** $x^2 + xy + y^2 - 8x + 8y = 0$

**d.** $xy = 5$

*Solution*

All of these parts have $B \neq 0$; therefore, we proceed as follows.

**a.** $B^2 - 4AC = 16 - 4(1)(4) = 0$; parabola

**b.** $B^2 - 4AC = 9 - 4(2)(1) > 0$; hyperbola

**c.** $B^2 - 4AC = 1 - 4(1)(1) < 0$; ellipse

**d.** $B^2 - 4AC = 1 - 4(0)(0) > 0$; hyperbola    □

### 5.3 The Rotation Concept

How can we do this?
To graph a conic that has been rotated, we need to introduce
a new coordinate axis for which the curve is in standard position.

This question can be answered by considering how you were able to read the preceding paragraph. What did you do? Probably you turned the book until the paragraph was right-side up. We do the same for a rotated conic.

If we rotated the axis through an angle $\theta$ ($0 < \theta < 90°$), the relationship between the old coordinates $(x, y)$ and the new coordinates $(x', y')$ can be found by considering Figure 44.

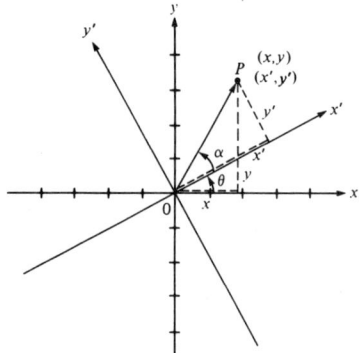

Figure 44  Rotation of axes

Let $O$ be the origin and $P$ be a point with coordinates $(x, y)$ relative to the old coordinate system and $(x', y')$ relative to the new rotated coordinate system. Let $\theta$ be the amount of rotation, and let $a$ be the angle between the $x'$-axis and $OP$. Then, using the definition of sine and cosine, we have

$$x = OP \cos(\theta + a)$$
$$y = OP \sin(\theta + a)$$

$$x' = OP \cos a$$
$$y' = OP \sin a$$

Using the identities of the sine and cosine of the sum of two angles we find:

**ROTATION FORMULAS**

$$x = x' \cos \theta - y' \sin \theta$$

$$y = x' \sin \theta + y' \cos \theta$$

### 5.4 What's Your Angle?

It is important that we rotate the axes "the right amount." That is, the new axes should be rotated the same amount as the given conic so that it will be in standard position after the rotation. To find out how much to rotate the axes, we substitute

$$x = x' \cos \theta - y' \sin \theta$$
$$y = x' \sin \theta + y' \cos \theta$$

into

$$Ax^2 + Bxy + Cy^2 + Dx + Ey + F = 0$$

We obtain (after a lot of simplifying)

$(A \cos^2 \theta + B \cos \theta \sin \theta + C \sin^2 \theta)x'^2$

$\qquad + [B(\cos^2 \theta - \sin^2 \theta) + 2(C - A)\sin \theta \cos \theta]x'y'$

$\qquad + (A \sin^2 \theta - B \sin \theta \cos \theta + \cos^2 \theta)y'^2$

$\qquad + (D \cos \theta + E \sin \theta)x' + (-D \sin \theta + E \cos \theta)y' + F = 0$

This looks terrible, but we want to choose $\theta$ so that there is no $x'y'$-term. That is

$$B(\cos^2 \theta - \sin^2 \theta) + 2(C - A)\sin \theta \cos \theta = 0$$

Simplifying, we obtain the following formula used for the angle of rotation. (Remember that $B \neq 0$ or there would have been no rotation in the first place.)

**ANGLE OF ROTATION**

$$\cot 2\theta = \frac{A - C}{B}$$

Find the appropriate rotation in Examples 4–6 so that the given curve will be in standard position relative to the rotated axes. Also find the $x$ and $y$ values in the new coordinate system.

**EXAMPLE 4**     $xy = 6$.

*Solution*

$$\cot 2\theta = \frac{A - C}{B}$$

$$= \frac{0 - 0}{1}$$

$$= 0$$

Thus, $2\theta = 90°$, $\theta = 45°$.

$$x = x' \cos \theta - y' \sin \theta$$

$$= x'\left(\frac{1}{\sqrt{2}}\right) - y'\left(\frac{1}{\sqrt{2}}\right) \qquad \cos 45° = \sin 45° = \frac{1}{\sqrt{2}}$$

$$= \frac{1}{\sqrt{2}}(x' - y')$$

$$y = x' \sin \theta + y' \cos \theta$$

$$= x'\left(\frac{1}{\sqrt{2}}\right) + y'\left(\frac{1}{\sqrt{2}}\right)$$

$$= \frac{1}{\sqrt{2}}(x' + y') \qquad \qquad □$$

**EXAMPLE 5**

$$7x^2 - 6\sqrt{3}xy + 13y^2 - 16 = 0.$$

*Solution*

$$\cot 2\theta = \frac{A - C}{B}$$

$$= \frac{7 - 13}{-6\sqrt{3}}$$

$$= \frac{1}{\sqrt{3}}$$

Thus, $2\theta = 60°$, $\theta = 30°$.

$$x = x' \cos \theta - y' \sin \theta \qquad y = x' \sin \theta + y' \cos \theta$$

$$= x'\left(\frac{\sqrt{3}}{2}\right) - y'\left(\frac{1}{2}\right) \qquad = x'\left(\frac{1}{2}\right) + y'\left(\frac{\sqrt{3}}{2}\right)$$

$$= \tfrac{1}{2}(\sqrt{3}x' - y') \qquad \qquad = \tfrac{1}{2}(x' + \sqrt{3}y') \qquad □$$

**EXAMPLE 6**

$$x^2 - 4xy + 4y^2 + 5\sqrt{5}y - 10 = 0.$$

*Solution*

$$\cot 2\theta = \frac{1 - 4}{-4} = \frac{3}{4}$$

This does not give us an exact value for $\theta$, so we need to use the following identities from trigonometry:

*First quadrant; choose positive value*

$$\downarrow$$

$$\tan 2\theta = \frac{1}{\cot 2\theta} = \frac{4}{3} \quad \text{and} \quad \sec 2\theta = \pm\sqrt{1 + \tan^2 2\theta} = \sqrt{1 + \tfrac{16}{9}} = \frac{5}{3}$$

This tells us that $\cos 2\theta = \frac{3}{5}$. Finally,

$$\cos \theta = \pm\sqrt{\frac{1 + \cos 2\theta}{2}} = \sqrt{\frac{1 + \frac{3}{5}}{2}} = \frac{2}{\sqrt{5}}$$

$$\sin \theta = \pm\sqrt{\frac{1 - \cos 2\theta}{2}} = \sqrt{\frac{1 - \frac{3}{5}}{2}} = \frac{1}{\sqrt{5}}$$

We choose the plus sign for cosine and sine because $\theta$ is in the first quadrant. Hence,

$$x = x' \cos \theta - y' \sin \theta \qquad y = x' \sin \theta + y' \cos \theta$$

$$= x'\left(\frac{2}{\sqrt{5}}\right) - y'\left(\frac{1}{\sqrt{5}}\right) \qquad = x'\left(\frac{1}{\sqrt{5}}\right) + y'\left(\frac{2}{\sqrt{5}}\right)$$

$$= \frac{1}{\sqrt{5}}(2x' - y') \qquad = \frac{1}{\sqrt{5}}(x' + 2y')$$

To find the rotation (or the slope of the rotated axes), we first find

$$m = \tan \theta = \frac{\sin \theta}{\cos \theta} = \frac{\frac{1}{\sqrt{5}}}{\frac{2}{\sqrt{5}}} = \frac{1}{2} \qquad \qquad \Box$$

Recall that the slope of the $x'$-axis is the tangent of the angle of inclination. Since $\theta$ is the angle of inclination, we draw the $x'$-axis so that it has a rise of one unit and a run of two units (from $m = \frac{1}{2}$). Notice that we found the rotated axes without ever consulting a calculator or a table of values. If you do want to know the angle of rotation in Example 6, use the inverse tangent function on your calculator to find $\theta \approx 26.6°$.

### 5.5 Examples of Rotated Conics

**PROCEDURE
TO SKETCH A
ROTATED CONIC**

**To sketch a rotated conic:**

1. Determine the nature of the conic by calculating $B^2 - 4AC$.
2. Find the angle of rotation.
3. Find $x$ and $y$ in the new coordinate system.
4. Substitute the values found in step 3 into the given equation, and simplify.
5. Sketch the resulting equation relative to the new $x'$-, $y'$-axes. You may have to complete the square if it is not centered at the origin.

**EXAMPLE 7**

Sketch $xy = 6$.

*Solution*

The graph of this equation is a hyperbola, since

$$B^2 - 4AC = 1 - 0 > 0.$$

From Example 4, $\theta = 45°$, and $x = \frac{1}{\sqrt{2}}(x' - y')$ and $y = \frac{1}{\sqrt{2}}(x' + y')$.

Substitute these values into the original equation:

$$\left[\frac{1}{\sqrt{2}}(x' - y')\right]\left[\frac{1}{\sqrt{2}}(x' + y')\right] = 6$$

Simplify (the details are not shown):

$$x'^2 - y'^2 = 12$$

$$\frac{x'^2}{12} - \frac{y'^2}{12} = 1$$

The sketch is shown in Figure 45.

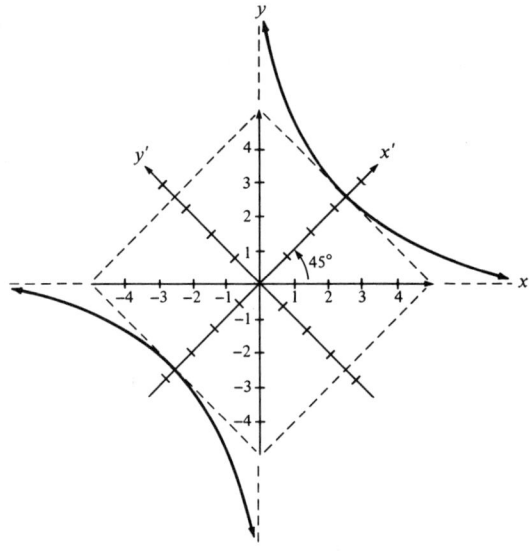

Figure 45  Graph of $xy = 6$

**EXAMPLE 8**          Sketch $7x^2 - 6\sqrt{3}xy + 13y^2 - 16 = 0$.

*Solution*          The graph of this equation is an ellipse, since

$$B^2 - 4AC = 36(3) - 4(7)(13) < 0$$

From Example 5, $\theta = 30°$, $x = \frac{1}{2}(\sqrt{3}x' - y')$, and $y = \frac{1}{2}(x' + \sqrt{3}y')$.

Substitute this into the original equation:

$$7(\tfrac{1}{2})(\sqrt{3}x' - y')^2 - 6\sqrt{3}(\tfrac{1}{2})(\sqrt{3}x' - y')(\tfrac{1}{2})(x' + \sqrt{3}y') + 13(\tfrac{1}{2})^2(x' + \sqrt{3}y')^2 - 16 = 0$$

Simplifying (again, not showing the steps), we obtain

$$\frac{x'^2}{4} + \frac{y'^2}{1} = 1$$

The sketch is shown in Figure 46.

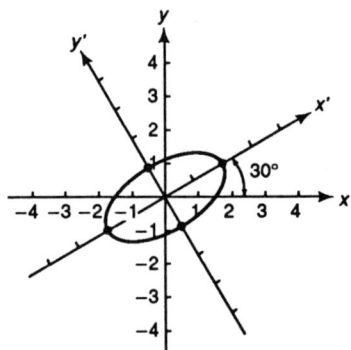

Figure 46  Graph of $7x^2 - 6\sqrt{3}xy + 13y^2 - 16 = 0$

**EXAMPLE 9**

Sketch $x^2 - 4xy + 4y^2 + 5\sqrt{5}y - 10 = 0$.

*Solution*

The graph of this equation is a parabola, since

$$B^2 - 4AC = 16 - 4(1)(4) = 0$$

From Example 6, the rotation is given by $\tan\theta = \frac{1}{2}$, and

$$x = \frac{1}{\sqrt{5}}(2x' - y') \quad \text{and} \quad y = \frac{1}{\sqrt{5}}(x' + 2y')$$

Substituting, we get

$$\frac{1}{5}(2x' - y')^2 - 4(\tfrac{1}{5})(2x' - y')(x' + 2y') + 4(\tfrac{1}{5})(x' + 2y')^2 + 5\sqrt{5}(\tfrac{1}{\sqrt{5}})(x' + 2y') - 10 = 0$$

Simplify (details not shown) to obtain

$$y'^2 + 2y' = -x' + 2$$

Complete the square to find

$$(y' + 1)^2 = -(x' - 3)$$

The sketch is shown in Figure 47.

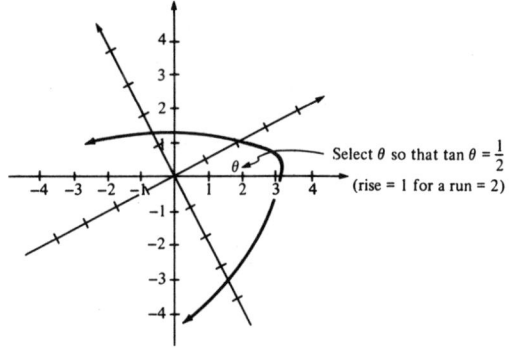

Select $\theta$ so that $\tan\theta = \frac{1}{2}$
(rise = 1 for a run = 2)

Figure 47  Graph of $x^2 - 4xy + 4y^2 + 5\sqrt{5}y - 10 = 0$  □

## 5.6 PROBLEM SET 5

**A**

1. Identify each of the curves given by the equations in Problems 3–9.

2. Identify each of the curves given by the equations in Problems 10–16.

*Find the appropriate rotation so that the curves given by the equations in Problems 3–16 will be in standard position relative to the rotated axes. Also find the x and y values in the new coordinate system.*

3. $xy = 8$

4. $xy = -4$

5. $xy = -1$

6. $13x^2 - 10xy + 13y^2 - 72 = 0$

7. $5x^2 - 26xy + 5y^2 + 72 = 0$

8. $x^2 + 4xy + 4y^2 + 10\sqrt{5}x = 9$

9. $5x^2 - 4xy + 8y^2 = 36$

10. $23x^2 + 26\sqrt{3}xy - 3y^2 - 144 = 0$

11. $3x^2 + 2\sqrt{3}xy + y^2 + 16x - 16\sqrt{3}y = 0$

12. $24x^2 + 16\sqrt{3}xy + 8y^2 - x + \sqrt{3}y - 8 = 0$

13. $3x^2 - 2\sqrt{3}xy + y^2 + 24x + 24\sqrt{3}y = 0$

14. $13x^2 - 6\sqrt{3}xy + 7y^2 + (16\sqrt{3} - 8)x + (-16 - 8\sqrt{3})y + 16 = 0$

15. $3x^2 - 10xy + 3y^2 - 32 = 0$

16. $5x^2 - 3xy + y^2 + 65x - 25y + 203 = 0$

*Sketch the curves in Problems 17–30.*

17. $xy = 8$

18. $xy = -4$

19. $xy = -1$

20. $13x^2 - 10xy + 13y^2 - 72 = 0$

21. $5x^2 - 26xy + 5y^2 + 72 = 0$

22. $x^2 + 4xy + 4y^2 + 10\sqrt{5}x = 9$

23. $5x^2 - 4xy + 8y^2 = 36$

24. $23x^2 + 26\sqrt{3}xy - 3y^2 - 144 = 0$

25. $3x^2 + 2\sqrt{3}xy + y^2 + 16x - 16\sqrt{3}y = 0$

26. $24x^2 + 16\sqrt{3}xy + 8y^2 - x + \sqrt{3}y - 8 = 0$

27. $3x^2 - 2\sqrt{3}xy + y^2 + 24x + 24\sqrt{3}y = 0$

28. $13x^2 - 6\sqrt{3}xy + 7y^2 + (16\sqrt{3} - 8)x + (-16 - 8\sqrt{3})y + 16 = 0$

29. $3x^2 - 10xy + 3y^2 - 32 = 0$

30. $5x^2 - 3xy + y^2 + 65x - 25y + 203 = 0$

# CONIC SECTIONS: SUMMARY

**CHAPTER 6**

## 6.1 Standard-form Equations

All conics can be reduced to standard-form equations by means of a translation or rotation. It is important to remember these equations and some basic information about the associated curves. The important ideas about each are summarized in Table 1.

**Table 1: CONIC SECTION SUMMARY**

| PARABOLA | ELLIPSE | HYPERBOLA |
|---|---|---|
| **Definition:**<br><br>All points equidistant from a given point and a given line. | All points the sum of whose distance from two fixed points is constant. | All points the difference of whose distance from two fixed points is constant. |
| **Equations:**<br><br>Opens up      $x^2 = 4cy$<br><br>Opens down    $x^2 = -4cy$<br><br>Opens right    $y^2 = 4cx$<br><br>Opens left     $y^2 = -4cx$ | $c^2 = a^2 - b^2$<br><br>Horizontal axis:<br><br>$\dfrac{x^2}{a^2} + \dfrac{y^2}{b^2} = 1$<br><br>Vertical axis:<br><br>$\dfrac{y^2}{a^2} + \dfrac{x^2}{b^2} = 1$ | $c^2 = a^2 + b^2$<br><br>Horizontal axis:<br><br>$\dfrac{x^2}{a^2} - \dfrac{y^2}{b^2} = 1$<br><br>Vertical axis:<br><br>$\dfrac{y^2}{a^2} - \dfrac{x^2}{b^2} = 1$ |
| **Recognition:**<br><br>Second-degree equation; linear in one variable, quadratic in the other variable. | Second degree equation; coefficients of $x^2$ and $y^2$ have same sign. | Second degree equation; coefficients of $x^2$ and $y^2$ have different signs. |
| **Distance from center to foci:** $c$<br>**Distance from center to vertex:** 0<br>**Eccentricity:**   $\epsilon = 1$<br>**Directrix:** one; $c$ units from center | $c$<br>$a$<br>$0 \le \epsilon < 1$<br>two; $\pm\frac{a}{\epsilon}$ units from center | $c$<br>$a$<br>$\epsilon > 1$<br>two; $\pm\frac{a}{\epsilon}$ units from center |

### 6.2 Transformation of Axes

The equations for the translation of axes to the point $(h, k)$ are

$$x' = x - h \quad \text{and} \quad y' = y - k.$$

The equations for the rotation of axes through an angle $\theta$ are

$$x = x' \cos \theta - y' \sin \theta \quad \text{and} \quad y = x' \sin \theta + y' \cos \theta$$

The amount of rotation $\theta$ for the curve

$$Ax^2 + Bxy + Cy^2 + Dx + Ey + F = 0$$

is found by

$$\cot 2\theta = \frac{A - C}{B}$$

---

*The String Method of Constructing Conics*

| *Circle* | *Parabola* | *Ellipse* | *Hyperbola* |
|---|---|---|---|
|  | 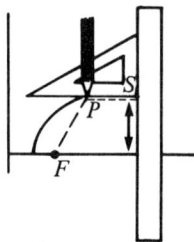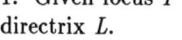 | 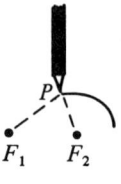 | 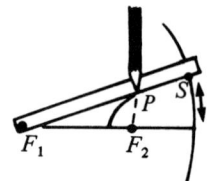 |
| 1. Given center $C$ and fixed radius $r$. | 1. Given focus $F$ and directrix $L$. | 1. Given foci $F_1$ and $F_2$ and fixed length $2a$. | 1. Given foci $F_1$ and $F_2$ and fixed length $2a$. |
| 2. Fix a string at $C$. How long should the string be? | 2. Fix a ruler parallel to $L$ at an arbitrary distance. | 2. Fix a string at $F_1$ and $F_2$. How long should the string be? | 2. A beam compass is set to describe an arbitrary circle with center at $F_1$. |
| 3. If the pencil keeps the string taut while tracing out a curve, the curve is a circle. | 3. Fix a string at $F$ and at a point $S$ on a triangle that moves along the ruler. How long does the string have to be? | 3. If the pencil keeps the string taut while tracing out a curve, the curve is an ellipse. | 3. A string is fastened to $F_2$ and a point $S$ on the beam. How long does the string have to be? |
| | 4. If the pencil keeps the string taut, the curve is a parabola. | | 4. If the pencil keeps the string taut while $S$ traces the circle, the curve is a hyperbola. |

### 6.3 Sketching a Conic

To graph the general second-degree equation

$$Ax^2 + Bxy + Cy^2 + Dx + Ey + F = 0$$

follow the steps shown in Figure 48.

*Step 1* Identify curve (this chart does not identify degenerate cases).

*Step 2* Rotate axis; if $B = 0$, go to step 3.

*Step 3* Translate axis.

*Step 4* Graph the standard-form equation.

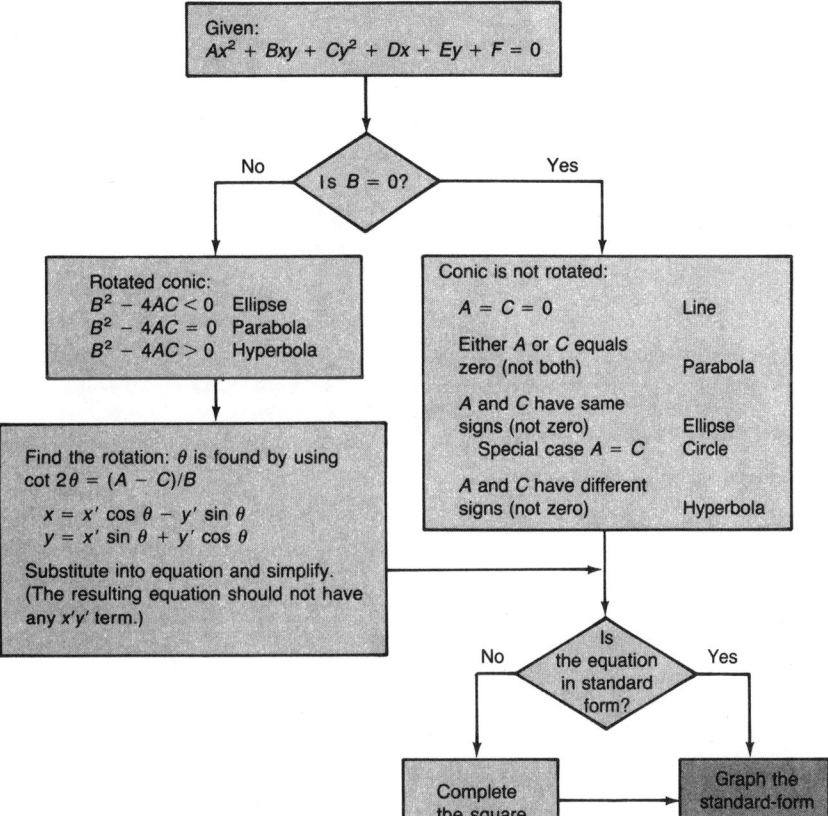

Figure 48 Procedure for graphing conics

### 6.4 PROBLEM SET 6

**A**

*In Problems 1–12, identify the curves and find the appropriate angle of rotation so that the given curve will be in standard position relative to the rotated axes. Also find the x and y values in the new coordinate system by using the rotation-of-axes formulas.*

1. $y^2 - 4x + 2y + 21 = 0$

2. $x^2 + 4x + 12y + 64 = 0$

3. $\dfrac{(x-3)^2}{4} - \dfrac{(y+2)}{6} = 1$

4. $x^2 + y^2 + 2x - 4y - 20 = 0$

5. $\dfrac{x}{9} + \dfrac{y}{25} = 1$

6. $x^2 - 4y^2 - 6x - 8y - 11 = 0$

7. $y^2 - 6y - 4x + 5 = 0$

8. $(x+3)^2 + (y-2)^2 = 0$

9. $9x^2 + 25y^2 - 54x - 200y + 256 = 0$

10. $xy = 4$

11. $x^2 + 2xy + y^2 + 12\sqrt{2}x - 6 = 0$

12. $8x^2 - 4xy + 5y^2 = 36$

*Graph the curves in Problems 13–24.*

13. $y^2 - 4x + 2y + 21 = 0$

14. $x^2 + 4x + 12y + 64 = 0$

15. $\dfrac{(x-3)^2}{4} - \dfrac{(y+2)}{6} = 1$

16. $x^2 + y^2 + 2x - 4y - 20 = 0$

17. $\dfrac{x}{9} + \dfrac{y}{25} = 1$

18. $x^2 - 4y^2 - 6x - 8y - 11 = 0$

19. $y^2 - 6y - 4x + 5 = 0$

20. $(x+3)^2 + (y-2)^2 = 0$

21. $9x^2 + 25y^2 - 54x - 200y + 256 = 0$

22. $xy = 4$

23. $x^2 + 2xy + y^2 + 12\sqrt{2}x - 6 = 0$

24. $8x^2 - 4xy + 5y^2 = 36$

*Sketch the graphs of Problems 25–35 on single coordinate axes. The result is a "picture" of a familiar object. The relation to the right of the colon is a limitation on the domain or the range.* [*]

25. $4x^2 + 16y = 0$:  $y \geq -4$

26. $2x - y - 10 = 0$:  $-14 \leq y \leq -12$

27. $2x - y - 8 = 0$:  $-13.6 \leq y \leq -12$

28. $x^2 + 256(y+12)^2 \leq 64$:  $2x - 8 \leq y \leq 2x - 10$

29. $4x^2 - 3y^2 - 24y - 112 = 0$:  $-12 \leq y \leq -4$

30. $16x^2 + 96x + 16y^2 + 480y + 3{,}708 = 0$

---

[*]These problems are from a design by Jan Schaafsma.

**31.** $x^2 + y^2 - 3y = 0$: $y \leq 2$

**32.** $100x^2 - 7y^2 + 98y - 368 = 0$:   $2 \leq y \leq 12$

**33.** $x^2 + 8(y - 12)^2 = 2$:   $y \geq 12$

**34.** $x^2 + 64(y + 4)^2 = 16$:   $y \geq -4$

**35.** $9x^2 + 2y^2 - 48y + 270 = 0$:   $y \geq 12$

# CHAPTER 7

# CURVE SKETCHING

### 7.1 A Picture Is Worth 1,000 Words

If a curve is not a conic, we must search out other methods of sketching it. These methods would, however, apply to the conics as well. Indeed, let us look at some of the main features that helped us sketch conics and apply them to higher-order curves.

When sketching a curve, we first check to see whether it is a type of curve we recognize (a line, a conic, a trigonometric function, or something else). If it is not a curve we recognize, we ultimately sketch it by plotting some points; but *before* we plot points we find out as much about the curve as we can. To do this we check (1) symmetry, (2) extent, (3) asymptotes, and (4) intercepts.

### 7.2 Symmetry

One of the most valuable tools in curve sketching is symmetry. This section deals with recognizing when a curve is symmetric by performing a simple test on its equation. In general, two points are symmetric with respect to a line if that line is the perpendicular bisector of the line containing the two points. In this course, we will check for symmetry with respect to the coordinate axes and the origin (see Figure 49).

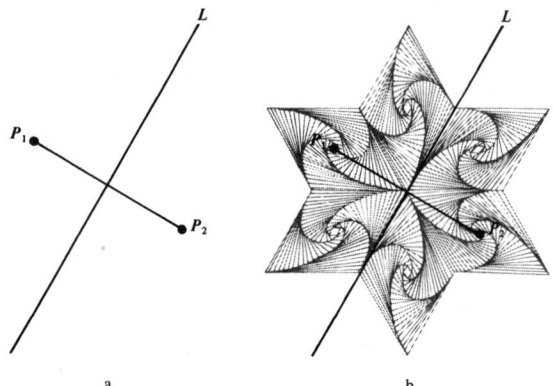

a                                        b

Figure 49     (a) Points $P_1$ and $P_2$ are symmetric with respect to $L$.
              (b) The star is symmetric with respect to $L$.

Figure 50 shows an example of a curve that is symmetric with respect to the $x$-axis. Whenever $(x,\ y)$ is on the curve, so is $(x,\ -y)$. This gives us a simple algebraic test: *if the equation remains unchanged when y is replaced by $-y$, then it is symmetric with respect to the x-axis,* because $(x,\ y)$ and $(x,\ -y)$ must both satisfy the equation.

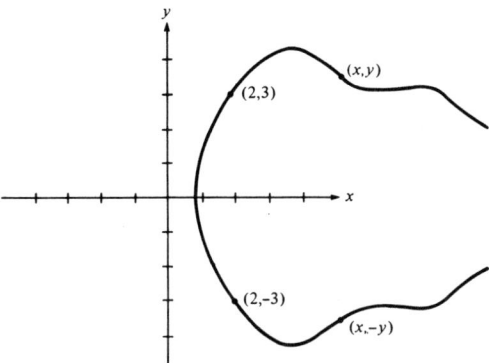

Figure 50 Symmetry with respect to the $x$-axis

Similarly, Figure 51 shows a curve that is symmetric with respect to the $y$-axis. Whenever $(x,\ y)$ is on the curve, so is $(-x,\ y)$. This means that, *if the equation is unchanged when x is replaced by $-x$, then it is symmetric with respect to the y-axis.*

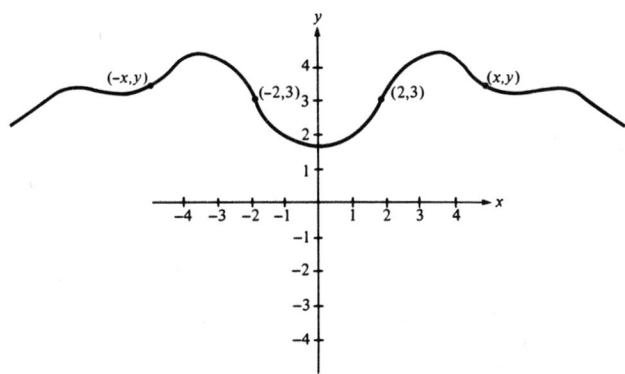

Figure 51 Symmetry with respect to the $y$-axis

Finally, Figure 52 shows a curve that is symmetric with respect to the origin. Whenever $(x,\ y)$ is on the curve, so is $(-x,\ -y)$. This means that, *if the equation is unchanged when x and y are replaced by $-x$ and $-y$, respectively, then it is symmetric with respect to the origin.*

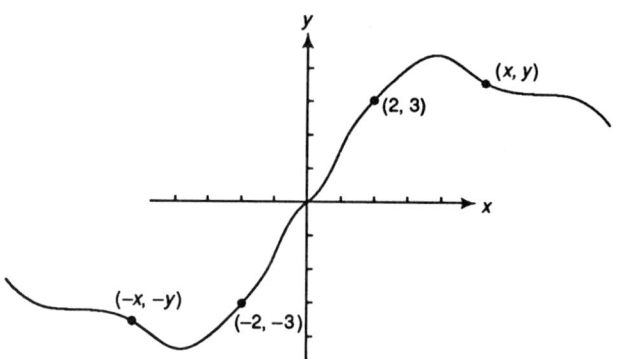

Figure 52  Symmetry with respect to the origin

**SYMMETRY**

A curve is symmetric with respect to

the **x-axis** if the equation remains unchanged when $y$ is replaced
by $-y$;
the **y-axis** if the equation remains unchanged when $x$ is
replaced by $-x$;
the **origin** if the equation remains unchanged when $x$ and $y$ are
simultaneously replaced by $-x$ and $-y$, respectively.
Also, the curve is symmetric with respect to the origin if
it is symmetric with respect to both the $x$- and $y$-axes.

**EXAMPLE 1**

*Solution*

Check the symmetry of $x^2 + 2xy^2 + y^2 = 4$.

This curve is symmetric with respect to the $x$-axis, since

$$x^2 + 2x(-y)^2 + (-y)^2 = 4$$

is the same as the original equation.  It is not symmetric with respect
to the $y$-axis or the origin.  □

**EXAMPLE 2**

*Solution*

Check the symmetry of $y = \cos x$.

This curve is symmetric with respect the the $y$-axis, since

$$y = \cos(-x)$$

is the same as the original equation.  It is not symmetric with respect
to the $x$-axis or the origin.  □

**EXAMPLE 3**

*Solution*

Check the symmetry of $x^3 + 2xy^2 + 4x^2y + 3y^2 = 0$.

This curve is symmetric with respect to the origin, since

$$(-x)^3 + 2(-x)(-y)^2 + 4(-x)^2(-y) = 0$$

is the same as the original equation if we multiply both sides by $-1$.
It is not symmetric with respect to the $x$- or $y$-axes.  □

**EXAMPLE 4**

*Solution*

Check the symmetry of $x^2 + 5x^2y^2 = 5$.

This curve is symmetric with respect to the $x$-axis, since

$$x^2 + 5x^2(-y)^2 = 5$$

is the same as the original equation. It is also symmetric with respect to the $y$-axis, since

$$(-x)^2 + 5(-x)^2y^2 = 5$$

is the same as the original equation. If a curve is symmetric with respect to both the $x$- and $y$-axes, then it must also be symmetric with respect to the origin. □

### 7.3 Extent

By *extent*, we mean the domain and range of a curve. If certain values of one or the other variable cause division by zero or imaginary values, those values must be excluded.

**PROCEDURE FOR FINDING THE EXTENT**

> **The domain is the set of all possible replacements for $x$. To find the domain:**
> 
> a. Solve for $y$ (if possible).
> b. The domain is the set of all real values for $x$ except those that
>    i. cause division by 0; or
>    ii. cause a negative under a square root (or other even-indexed root).
> 
> **The range is the set of all possible replacements for $y$. To find the range:**
> 
> a. Solve for $x$ (if possible).
> b. The range is the set of all real values for $y$ except those that
>    i. cause division by 0; or
>    ii. cause a negative under a square root (or other even-indexed root).

**EXAMPLE 5**

*Solution*

Find the domain and range of $y = \frac{4}{x}$.

The domain is the set of all real numbers *except* 0, since this value causes division by zero. In such a case, we write all reals, $x \neq 0$. For the range, solve for $x$:

$$x = \frac{4}{y}$$

The range is also the set of all real numbers, $y \neq 0$. □

**EXAMPLE 6**

*Solution*

Find the domain of $y = \dfrac{(x - 3)(x - 2)(x - 5)}{(x + 2)(x - 4)}$.

The domain is the set of all real numbers, $x \neq -2$ and $x \neq 4$. □

**EXAMPLE 7**

Find the domain and range of $y = \sqrt{\dfrac{x}{x-1}}$.

*Solution*

For the domain:

$$x \neq 1 \qquad \textit{Exclude values that cause division by 0.}$$

$$\frac{x}{x-1} \geq 0 \qquad \textit{Numbers under a square root must be non-negative.}$$

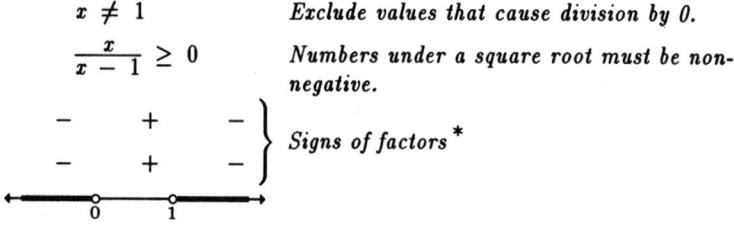

$\textit{Signs of factors}$ *

Domain: $(-\infty, 0) \cup (1, \infty)$

For the range, we solve for $x$:

$$y^2 = \frac{x}{x-1}$$

$$y^2(x-1) = x$$

$$y^2 x - y^2 = x$$

$$y^2 x - x = y^2$$

$$(y^2 - 1)x = y^2$$

$$x = \frac{y^2}{y^2 - 1}$$

We must exclude values that cause division by 0:

$$y^2 - 1 \neq 0$$

$$y \neq 1, -1$$

But $y$ cannot be negative, since $y$ equals a radical and radicals are non-negative. Therefore, the range is the set of all non-negative real numbers except $y = 1$.  □

## 7.4 Asymptotes

An asymptote is a line such that, as a point $P$ on the curve moves farther away from the origin, the distance between $P$ and the asymptote tends toward 0. In this section, we are concerned with finding horizontal, vertical, and slant asymptotes. If we set $x$ and $y$ equal to constants that cause division by 0, we will have the equations of the horizontal and vertical asymptotes.

Sometimes it is not convenient (or possible) to solve an equation for $x$. Using calculus, it can be shown that, if

$$y = \frac{P(x)}{D(x)}$$

where $P(x)$ and $D(x)$ are polynomial functions of $x$ with no common factors (that is, where the rational expression is reduced), the asymptotes depend on the degree of $P$ and $D$. Suppose that $P(x)$ has degree $M$ with a leading coefficient $p$, and that $D(x)$ has degree $N$ with

---

*See Appendix B for a review of how to solve inequalities.

leading coefficient *d*. The asymptotes can then be found according to the rules given in the following box.

**PROCEDURE FOR FINDING ASYMPTOTES**

**Vertical asymptotes:** $x = r$, where $r$ is a value that causes division by 0 when the equation is solved for $y$ and is reduced.

**Horizontal asymptotes:** $y = r$, where $r$ is a value that causes division by 0 when the equation is solved for $x$ and is reduced. Sometimes it is not possible (or convenient) to solve for $x$. If the equation is solved for $y$ and is reduced, then

$y = 0$ is a horizontal asymptote if $M < N$

$y = \frac{p}{d}$ is a horizontal asymptote if $M = N$

no horizontal asymptote exists if $M > N$

**Slant asymptotes:** Solve the equation for $y$ and reduce; then

$y = mx + b$ is a slant asymptote if $M = N + 1$

where $mx + b$ is the quotient (without remainder) obtained when $P(x)$ is divided by $D(x)$.

In Example 5, there is a vertical asymptote, $x = 0$, and a horizontal asymptote, $y = 0$. In Example 6, there are two vertical asymptotes, $x = -2$ and $x = 4$, and a horizontal asymptote, $y = 1$. In Example 7, there is a vertical asymptote, $x = 1$, and a horizontal asymptote, $y = 1$.

**EXAMPLE 8**

Find the vertical, horizontal, and slant asymptotes for

$$y = \frac{6x^2 - x - 1}{4x^2 - 4x + 1}$$

*Solution*

First, make sure the rational function is reduced:

$$y = \frac{6x^2 - x - 1}{4x^2 - 4x + 1}$$

$$= \frac{(3x + 1)(2x - 1)}{(2x - 1)^2}$$

$$= \frac{3x + 1}{2x - 1}$$

**Vertical asymptote:** can be found when $2x - 1 = 0$; $x = \frac{1}{2}$ is the equation of a vertical asymptote.

**Horizontal asymptote:** $y = \frac{3}{2}$, which is found by looking at the leading coefficients of the reduced form.

**Slant asymptotes:** do not exist for this curve, since the degree of the numerator is not one more than the degree of the denominator. □

**EXAMPLE 9**

Find the vertical, horizontal, and slant asymptotes for

$$y = \frac{3x^2 - 2x^2 - 9x + 6}{x^2 - 3}$$

*Solution*

The expression is reduced. If you use long division,

$$
\begin{array}{r}
3x - 2 \\
x^2 - 3 \overline{\smash{\big)}\ 3x^3 - 2x^2 - 4x + 16} \\
\underline{3x^3 \qquad\quad - 9x} \\
-2x^2 + 5x + 16 \\
\underline{-2x^2 \qquad + 6} \\
5x + 10
\end{array}
$$

Since there is a remainder, the rational expression is reduced.

**Vertical asymptotes:** can be found when $x^2 - 3 = 0$; $x = \pm\sqrt{3}$.

**Horizontal asymptotes:** none, since the degree of the numerator is larger than the degree of the denominator.

**Slant asymptotes:** exist when the degree of the numerator is one more than the degree of the denominator. Carry out the long division (as shown above) and disregard the remainder; the slant asymptote is

$$y = 3x - 2 \qquad\qquad\qquad \square$$

### 7.5  Intercepts

The intercepts are the places where the curve crosses a coordinate axis. When finding the intercepts, we are really plotting points on the curve, but these are generally the easiest points to find.

**PROCEDURE FOR
FINDING INTERCEPTS**

> **x-intercepts:**  set $y = 0$ and solve for $x$ (if possible)
> **y-intercepts:**  set $x = 0$ and solve for $y$ (if possible)

### 7.6  Examples of Curve Sketching

**EXAMPLE  10**

Sketch the curve $x^2(y - 2) = 2$, given the following information:

> Symmetry: with respect to the $y$-axis
> Domain: $x \neq 0$
> Range: $y > 2$
> Asymptotes: $x = 0$, $y = 2$
> Intercepts: none

*Solution*

Since we have symmetry with respect to the $y$-axis, we focus our attention on Quadrants I and IV. We draw the asymptotes and limit our attention to the domain and range. It is customary to darken the parts of the plane **not** included in the domain or range, as shown by the shaded portion of Figure 53(a). Since there are no intercepts, we plot the following points:

$$x = 1, \text{ then } 1(y - 2) = 2, \text{ so } y = 4$$
$$x = 2, \text{ then } 4(y - 2) = 2, \text{ so } y = \tfrac{5}{2}$$

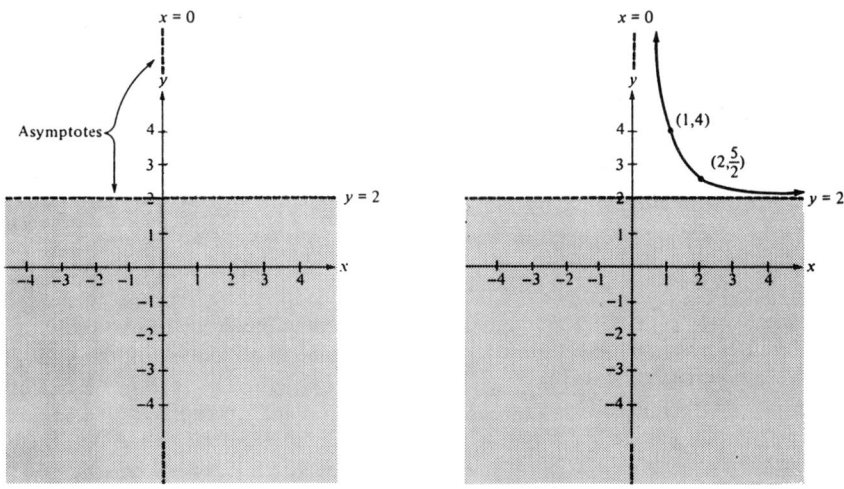

Figure 53

We plot these points and, using the fact that $x = 0$ and $y = 2$ are asymptotes, sketch the part of the curve in the first quadrant, as shown in Figure 53(b). By symmetry, we then sketch the rest of the curve, as shown in Figure 54.

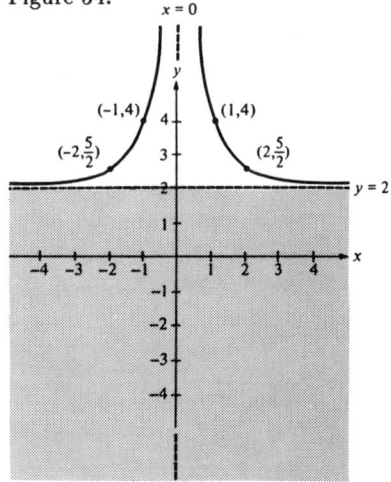

Figure 54  Graph of $x^2(y - 2) = 2$ ☐

**EXAMPLE 11**  Sketch $x^2 = \dfrac{1 + y^2}{1 - y^2}$.

*Solution*  **Symmetry:**  The curve is symmetric with respect to the $x$-axis, since

$$x^2 = \frac{1 + (-y)^2}{1 - (-y)^2}$$

is the same as the original equation; it is also symmetric with respect to the $y$-axis and the origin.

**Extent:**    **domain:** Solve for $y$.

$$x^2 = \frac{1 + y^2}{1 - y^2}$$

$$x^2 - x^2 y^2 = 1 + y^2$$

$$x^2 - 1 = y^2 + x^2 + x^2 y^2$$

$$x^2 - 1 = (1 + x^2) y^2$$

$$y^2 = \frac{x^2 + 1}{x^2 - 1}$$

$$y = \pm \sqrt{\frac{x^2 - 1}{x^2 + 1}}$$

We must now rule out values of $x$ that cause division by 0 or negative values under the square root radical. Solve

$$\frac{x^2 - 1}{x^2 + 1} \geq 0$$

Appendix B discusses solving quadratic inequalities. The solution for this inequality gives the domain:

$$x \leq -1 \quad \text{or} \quad x \geq 1$$

**range:** Solve for $x$.

$$x = \pm \sqrt{\frac{1 + y^2}{1 - y^2}}$$

We need to find the $y$-values that cause division by 0 or negative values under the square root radical. First solve

$$1 - y^2 = 0$$

to find $y = \pm 1$. These are *excluded values*. Next solve

$$\frac{1 + y^2}{1 - y^2} \geq 0, \quad y \neq \pm 1$$

to find the range:  $-1 < y < 1$

Use the information about extent to darken (shade) the portions of the plane that *cannot* contain the graph. This is shown in Figure 55.

**Asymptotes:** **vertical:**

$$y = \pm \frac{\sqrt{x^2 - 1}}{\sqrt{x^2 + 1}} \qquad \textit{No division by 0, so there are no vertical asymptotes.}$$

**horizontal:**

$$x = \pm \frac{\sqrt{1 + y^2}}{\sqrt{1 - y^2}} \qquad \textit{Division by 0 when } y = \pm 1.$$

Horizontal asymptotes are $y = 1$, $y = -1$; draw these as dashed lines on the coordinate axes.

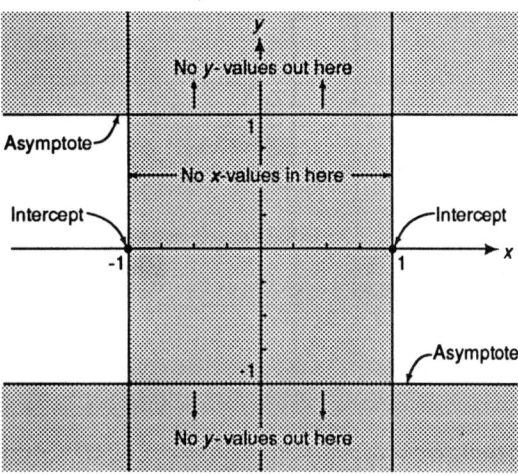

Figure 55

**Intercepts:**    **$x$-intercepts:** if $y = 0$, then $x = \pm 1$; so the $x$-intercepts are at $(1, 0)$ and $(-1, 0)$.

               **$y$-intercepts:** if $y = \pm\sqrt{-1}$; so there are no $y$-intercepts.

**Plot points:**  Plot one or more relevant points, and make use of the information obtained above to sketch the curve, as shown in Figure 56.

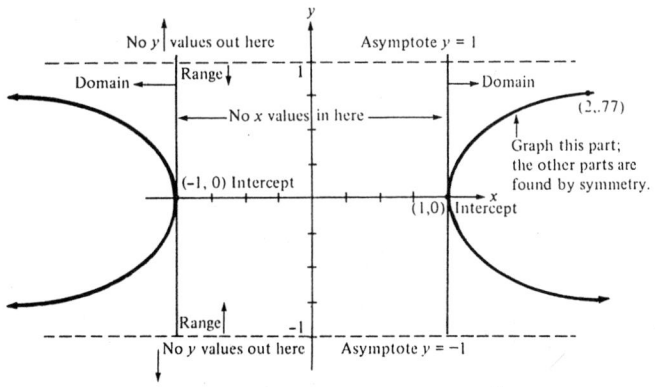

Figure 56  Graph of $x^2 = \dfrac{1 + y^2}{1 - y^2}$         □

**EXAMPLE 12**

Sketch $xy + 4y = 1$.

*Solution*

We can sometimes combine the methods of this chapter with the methods of Chapter 5. That is, we recognize this curve as a rotated conic. Indeed, since $B^2 - 4AC > 0$, it is a rotated hyperbola.

Also, $\cot 2\theta = \dfrac{A - C}{B} = 0$, which means that $\theta = 45°$. Next, however, instead of proceeding as in Chapter 5, we use some of the information of this chapter.

**Symmetry:** none

**Extent:**   **domain:** Since $y = \dfrac{1}{x + 4}$, the domain is the set of all real numbers, except $x = -4$

**range:**   Since $x = \dfrac{1 - 4y}{y}$, the range is the set of all real numbers, except $x = 0$.

**Asymptotes:** $x = -4$ and $y = 0$ are the asymptotes.

**Intercepts:**   **$x$-intercept:** none (since $y \neq 0$);

**$y$-intercept:** $(0, \frac{1}{4})$

**Plot points** and make use of the information above to sketch the curve, as shown in Figure 57.

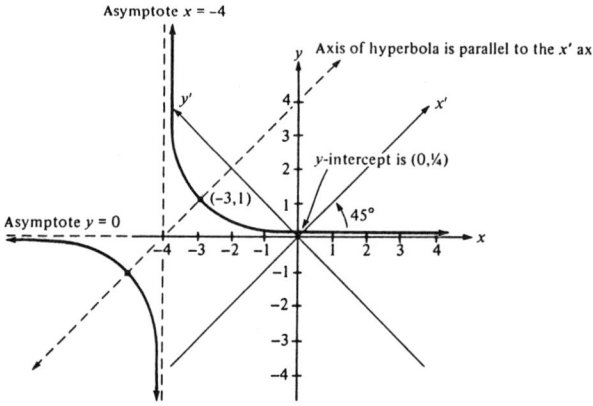

**Figure 57**   Graph of $xy + 4y = 1$                            ☐

### 7.7  PROBLEM SET 7

**A**

*In Problems 1–4, use the given information and plot some points to draw a sketch of the curve.*

1.  $x^2y = 4$
    Symmetry: with respect to the *y*-axis
    Domain: $x \neq 0$
    Range: $y > 0$
    Asymptotes: $x = 0$; $y = 0$
    Intercepts: none

2.  $xy^2 - y^2 - 1 = 0$
    Symmetry: with respect to the *x*-axis
    Domain: $x > 1$
    Range: $y \neq 0$
    Asymptotes: $x = 1$; $y = 0$
    Intercepts: none

3.  $x^2y - 4x + 2y = 0$
    Symmetry: with respect to the origin
    Domain: all real numbers
    Range: $(-\sqrt{2}, \sqrt{2})$
    Asymptotes: $y = 0$
    Intercept: $(0, 0)$

4.  $x^2y^2 - 4xy^2 + 3y^2 - 4 = 0$
    Symmetry: with respect to the *x*-axis
    Domain: $(-\infty, 1) \cup (3, \infty)$
    Range: $y \neq 0$
    Asymptotes: $x = 1$; $x = 3$; $y = 0$
    Intercepts: $(0, \frac{2}{3}\sqrt{3})$, $(0, -\frac{2}{3}\sqrt{3})$

*Find the symmetry, extent, asymptotes, and intercepts for the curves whose equations are given in problems 5–20.*

5.  $xy = 2$

6.  $xy = 6$

7.  $y = \dfrac{x + 1}{x}$

8.  $y = \dfrac{x + 1}{x + 2}$

9.  $y = \dfrac{2x^2 + x - 10}{x + 2}$

10. $y = \dfrac{3x^2 + 5x - 2}{x + 2}$

11. $y = \dfrac{2x^3 - 3x^2 - 2x}{2x + 1}$

12. $y = \dfrac{x^3 + 6x^2 + 15x + 14}{x + 2}$

13. $9x^2 + 4y^2 - 36 = 0$

14. $6x^2 - 2y^2 + 10 = 0$

15. $13x^2 - 10xy + 13y^2 - 72 = 0$

16. $y^2x - 2y^2 + 2 = 0$

17. $x^2y - 4xy + 3y - 4 = 0$

18. $x^3 - y^2 - 4y = 0$

19. $x^4 - x^2y^2 - 4x^2 + y^2 = 0$

20. $2y^2 - xy^2 + x - 1 = 0$

*Graph the curves in Problems 21–36.*

21. $xy = 2$

22. $xy = 6$

23. $y = \dfrac{x+1}{x}$

24. $y = \dfrac{x+1}{x+2}$

25. $y = \dfrac{2x^2 + x - 10}{x+2}$

26. $y = \dfrac{3x^2 + 5x - 2}{x+2}$

27. $y = \dfrac{2x^3 - 3x^2 - 2x}{2x+1}$

28. $y = \dfrac{x^3 + 6x^2 + 15x + 14}{x+2}$

29. $9x^2 + 4y^2 - 36 = 0$

30. $6x^2 - 2y^2 + 10 = 0$

31. $13x^2 - 10xy + 13y^2 - 72 = 0$

32. $y^2x - 2y^2 + 2 = 0$

33. $x^2y - 4xy + 3y - 4 = 0$

34. $x^3 - y^2 - 4y = 0$

35. $x^4 - x^2y^2 - 4x^2 + y^2 = 0$

36. $2y^2 - xy^2 + x - 1 = 0$

# POLAR CURVES

**CHAPTER 8**

## 8.1 Plotting Points

Up to this point in the book, we have plotted points by using rectangular coordinates — that is, by measuring out a distance on the $x$-axis and then a distance on the $y$-axis to plot the point $(x, y)$. Now we consider a different method of locating points in the Cartesian plane. In this system, we measure a rotation of the $x$-axis (a positive or negative rotation $\theta$) and then measure a (positive or negative) distance $r$ along the rotated $x$-axis. A point fixed in this fashion is denoted by $P(r, \theta)$ and is called a **polar coordinate.** The point $(0, 0)$ is called the **origin** when rectangular coordinates are used and the **pole** when polar coordinates are used.

When plotting points in polar form, consider a **polar axis,** fixed at the pole and coinciding with the $x$-axis. Now rotate the polar axis through an angle $\theta$, as shown in Figure 58.

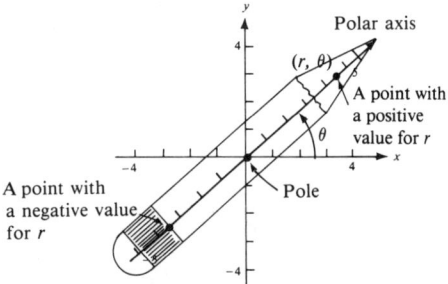

Figure 58  Plotting polar-form points

You might find it helpful to rotate your pencil as the axis; the tip points in the positive direction. If $\theta$ is positive, the angle is measured in a counterclockwise direction, and if $\theta$ is negative, it is measured in a clockwise direction. Next, plot $r$ on the polar axis (the pencil).

**EXAMPLE 1**

Plot each of the following polar-form points:

$A(4, \frac{\pi}{3})$; $B(-4, \frac{\pi}{3})$; $C(3, -\frac{\pi}{6})$; $D(-3, -\frac{\pi}{6})$; $E(-3, 3)$; $F(-3, -3)$;

---

*Optional chapter

73

$G(-4, -2)$; $H(5, \frac{3\pi}{2})$; $I(-5, \frac{\pi}{2})$; $J(5, -\frac{\pi}{2})^*$

*Solution*    Points $A$, $B$, $C$, and $D$ illustrate basic polar-form plotting.

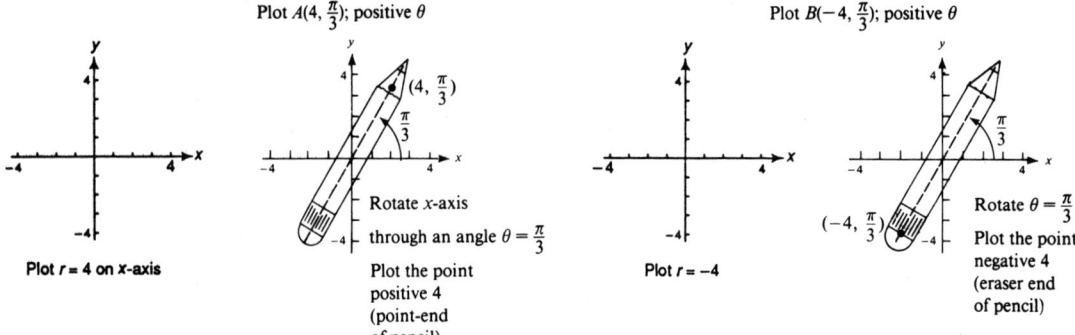

Points $E$, $F$, $G$, and $H$ illustrate situations that can sometimes be confusing.  Make sure you understand each part of this example.

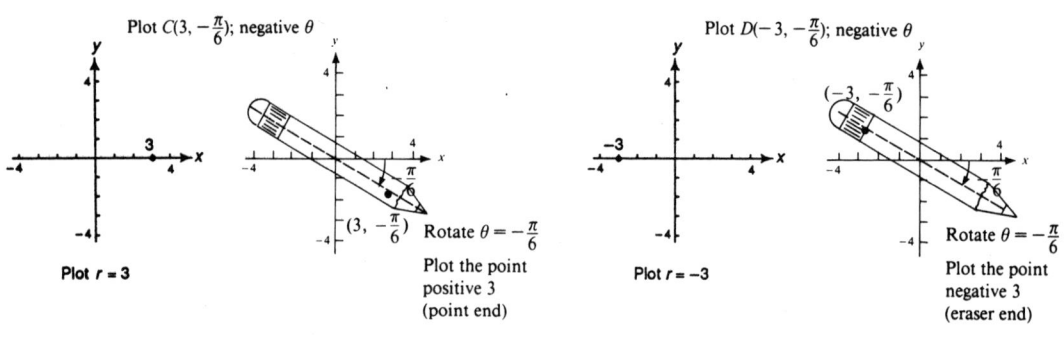

## 8.2  Primary Representations of a Point

One thing you will notice from Example 1 is that ordered pairs in polar form are not associated in a one-to-one fashion with points in the plane.  Indeed, given any point in the plane, infinitely many ordered pairs of polar coordinates are associated with that point in polar form.  If you are given a point $(r, \theta)$ other than the pole in polar coordinates, then $(-r, \theta + \pi)$ also represents the same point.  In addition, there are infinitely many others, all of which have the same first component as one of these and second components that are multiples of $2\pi$ added to these angles.  We call $(r, \theta)$ and $(-r, \theta + \pi)$ the **primary representations of the point** if the angles $\theta$ and $\theta + \pi$ are between 0 and $2\pi$.

**PRIMARY
REPRESENTATIONS**

> Every point in polar form has two primary representations:
> $(r, \theta)$, where $0 \le \theta < 2\pi$, and $(-r, \pi + \theta)$, where $0 \le \pi + \theta < 2\pi$

---

*Angle measure with units is understood to be radian measure.  The real number $2\pi$ will trace out an angle of one revolution.

**EXAMPLE 2**

Give both primary representations for each of the given points.

**a.** $(3, \frac{\pi}{4})$ has primary representations $(3, \frac{\pi}{4})$ and $(-3, \frac{5\pi}{4})$.
$$\uparrow$$
$$\frac{\pi}{4} + \pi = \frac{5\pi}{4}$$

**b.** $(5, \frac{5\pi}{4})$ has primary representations $(5, \frac{5\pi}{4})$ and $(-5, \frac{\pi}{4})$.
$$\uparrow$$

> $\frac{5\pi}{4} + \pi = \frac{9\pi}{4}$, *but* $(-5, \frac{9\pi}{4})$ *is not a primary representation since* $\frac{5\pi}{4} > 2\pi$. *Use* $\frac{\pi}{4}$, *since it is coterminal with* $\frac{9\pi}{4}$ *and satisfies* $0 \leq \theta < 2\pi$.

**c.** $(-6, -\frac{2\pi}{3})$ has primary representations $(-6, \frac{4\pi}{3})$ and $(6, \frac{\pi}{3})$.

**d.** $(9, 5)$ has primary representations $(9, 5)$ and $(-9, 5 - \pi)$.
$$\uparrow$$

> *A point like* $(-9, 5 - \pi)$ *is usually approximated by writing* $(-9, 1.86)$. *Also notice that* $(-9, 5 + \pi)$ *is not a primary representation, since* $5 + \pi > 2\pi$.

**e.** $(9, 7)$ has primary representation $(9, .72)$ and $(-9, 3.86)$.
$$\qquad\qquad\qquad\uparrow\qquad\qquad\qquad\uparrow$$
$$7 - 2\pi \approx .72 \qquad\qquad 7 - \pi \approx 3.86 \quad \square$$

### 8.3 Relationship Between Polar and Rectangular Coordinates

The relationship between rectangular and polar coordinates can easily be found by using the definitions of the trigonometric functions (see Figure 59).

**RELATIONS BETWEEN RECTANGULAR AND POLAR COORDINATES**

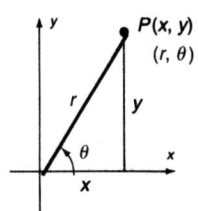

Figure 59

1. To change from polar to rectangular coordinates:
$$x = r \cos \theta \qquad\qquad y = r \sin \theta$$

2. To change from rectangular to polar coordinates:
$$r = \sqrt{x^2 + y^2} \qquad\qquad \theta' = \tan^{-1}\left|\frac{y}{x}\right|, \ x \neq 0$$
where $\theta'$ is the reference angle for $\theta$. Place $\theta$ in the proper quadrant by noting the signs of $x$ and $y$. If $x = 0$, then $\theta' = \frac{\pi}{2}$.

Remember from trigonometry that the notation $\tan^{-1} t$ means inverse tangent (or angle whose tangent is $t$) and does *not* mean $1/\tan t$.

**EXAMPLE 3**

Change the polar coordinates $(-3, \frac{5\pi}{4})$ to rectangular coordinates.

*Solution*

$$x = -3 \cos \frac{5\pi}{4} = -3\left(-\frac{\sqrt{2}}{2}\right) = \frac{3\sqrt{2}}{2}$$
$$y = -3 \sin \frac{5\pi}{4} = -3\left(-\frac{\sqrt{2}}{2}\right) = \frac{3\sqrt{2}}{2}$$

$$\underbrace{\left(-3, \frac{5\pi}{4}\right)}_{\textit{polar form}} = \underbrace{\left(\frac{3\sqrt{2}}{2}, \frac{3\sqrt{2}}{2}\right)}_{\textit{rectangular form}}$$

$\qquad\qquad\qquad\qquad\qquad\qquad\qquad\qquad\qquad\qquad\qquad\qquad\square$

**EXAMPLE 4**

Write both primary representations of the polar-form coordinates for the point whose rectangular coordinates are $\left(\frac{5\sqrt{3}}{2}, -\frac{5}{2}\right)$

*Solution*

$$r = \sqrt{\left(\frac{5\sqrt{3}}{2}\right)^2 + \left(-\frac{5}{2}\right)^2} = \sqrt{\frac{75}{4} + \frac{25}{4}} = 5$$

$$\theta' = \tan^{-1}\left|\frac{-\frac{5}{2}}{\frac{5\sqrt{3}}{2}}\right| = \tan^{-1}\left(\frac{1}{\sqrt{3}}\right) = \frac{\pi}{6}; \; \theta = \frac{11\pi}{6} \quad \textit{Quadrant IV}$$

$$\underbrace{\left(\frac{5\sqrt{3}}{2}, -\frac{5}{2}\right)}_{\text{rectangular form}} = \underbrace{\left(5, \frac{11\pi}{6}\right) = \left(-5, \frac{5\pi}{6}\right)}_{\text{polar form}}$$

□

## 8.4 Points on a Polar Curve

We now turn our attention to polar-form graphing. The basic procedure is to plot points to determine a curve's general shape and then to make some generalizations that will satisfy the graphing of similar curves. Because the representation of points in polar form as ordered pairs of real numbers is not unique, we need the following definition of what it means for a polar-form point to **satisfy** an equation.

**A POINT SATISFYING AN EQUATION**

An ordered pair representing a polar-form point (other than the pole) **satisfies an equation** involving a trigonometric function of $n\theta$ (where $n$ is an integer) if and only if at least one of its primary representations satisfies the given equation.

The easiest graphs to consider are those for which either $r$ or $\theta$ is a constant. For example, $r = 5$ is the equation of the set of all ordered pairs $(r, \theta)$ for which the first component, $r$, is 5. This means that the points $(5, 0)$, $(5, 1)$, $(5, 2)$ , $(5, 3)$, $\cdots$ all satisfy this equation. The graph shown in Figure 60(a) appears as a circle with center at the pole and radius 5. Similarly, $\theta = 1$ is the set of all ordered pairs for which the second component, $\theta$, is 1. This means that the points $(-1, 1)$, $(0, 1)$, $(1, 1)$, $(2, 1)$, $(3, 1)$, $\cdots$ all satisfy the equation $\theta = 1$. The graph shown in Figure 60(b) appears as a line.

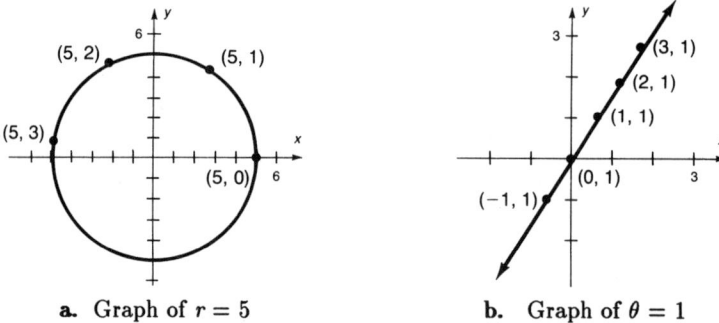

**a.** Graph of $r = 5$        **b.** Graph of $\theta = 1$

Figure 60  Polar-form graphs for which one component is a constant

The following equations show some nontrivial polar graphs. Remember, the basic ideas of polar-form graphing are identical to those of rectangular-form graphing. You are looking for ordered pairs satisfying the given equation. Each of these ordered pairs is then plotted on a Cartesian coordinate system; the only difference with polar-form graphing is that the first component represents a distance from the pole and the second component represents an angle of rotation.

### 8.5 Cardioids

**EXAMPLE 5**

Graph $r = 2(1 - \cos \theta)$.

*Solution*

First construct a table of values by choosing values for $\theta$ and approximating the corresponding values for $r$.

| $\theta$: | 0 | 1 | 2 | 3 | 4 | 5 | 6 | 7 |
|---|---|---|---|---|---|---|---|---|
| $r$: | 0 | .92 | 2.83 | 3.98 | 3.31 | 1.43 | .08 | .49 |

↑

*Be sure your calculator is set to radian units and press*
1 cos +/- + 1 = × 2 =

The graph is shown in Figure 61.

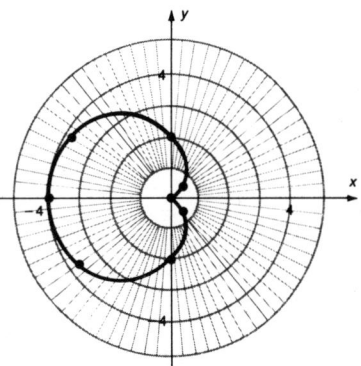

Figure 61  Graph of $r = 2(1 - \cos \theta)$, by plotting points  ☐

The graph in Example 5 is called a **cardioid** because it is heart-shaped. Compare the curve graphed in Example 5 with the general curve

$$r = a(1 - \cos \theta)$$

which is called the **standard-position cardioid**. Consider the following table of values:

| $\theta$: | 0 | $\frac{\pi}{2}$ | $\pi$ | $\frac{3\pi}{2}$ |
|---|---|---|---|---|
| $r$: | 0 | $a$ | $2a$ | $a$ |

These reference points are all that is necessary to plot cardioids, because they will all have the same shape as the one shown in Figure 61. For that cardioid you would simply plot the points $(0, 0)$, $(2, \frac{\pi}{2})$,

$(4, \pi)$, and $(2, \frac{3\pi}{2})$, and then draw the rest of the cardioid by remembering its shape.

### 8.6  Rotation of Polar-form Graphs

What about cardioids that are not in standard position? We have translated and rotated the conic sections, and we found translations to be easy and rotations difficult.  For polar-form curves, however, translations are difficult and rotations are easy.

---

**ROTATION OF POLAR-FORM GRAPHS**

> The polar graph $r = f(\theta - \alpha)$ is the same as the polar graph of $r = f(\theta)$ that has been rotated through an angle $\alpha$.

**EXAMPLE 6**

Graph $r = 3 - 3\cos(\theta - \frac{\pi}{6})$.

*Solution*

Recognize the curve as a cardioid with $a = 3$ and a rotation of $\frac{\pi}{6}$.  Plot the four points as shown in Figure 62, and draw the cardioid.

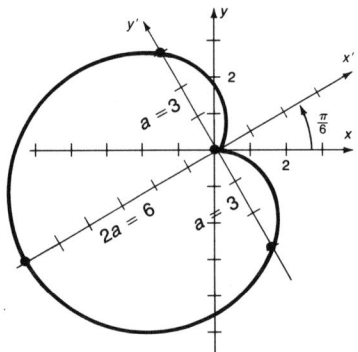

Notice that the $x'y'$-axis is drawn by rotating the $xy$-axis through an angle of $\frac{\pi}{6}$. The standard cardioid is then drawn on the rotated axis.

Figure 62  Graph of $r = 3 - 3\cos(\theta - \frac{\pi}{6})$    □

**EXAMPLE 7**

Graph $r = 4(1 - \sin\theta)$.

*Solution*

Notice that
$$\sin\theta = \cos(\frac{\pi}{2} - \theta) = \cos(\theta - \frac{\pi}{2})$$

Thus, $r = 4[1 - \cos(\theta - \frac{\pi}{2})]$. This is a cardioid with $a = 4$ that has been rotated $\frac{\pi}{2}$. The graph is shown in Figure 63. A curve of the form $r = a(1 - \sin\theta)$ is a cardioid that has been rotated $\frac{\pi}{2}$.

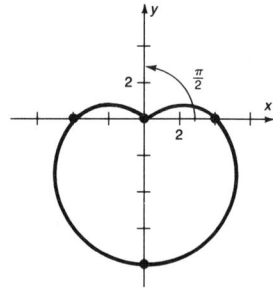

Figure 63  Graph of $r = 4(1 - \sin\theta)$    □

The cardioid is only one of the interesting polar-form curves.  It is a special case of a curve called a **limaçon**, which is developed in the problem set.

The next example illustrates a curve called a **rose curve**.

### 8.7 Rose Curves

**EXAMPLE 8**

Graph $r = 4 \cos 2\theta$.

*Solution*

When presented with a polar-form curve that you do not recognize, graph the curve by plotting points. You can use tables, exact values, or a calculator. A calculator was used to find the following values:

| $\theta$: | 0 | .25 | .5 | .75 | 1 | 1.25 | 1.5 | 2 | 2.5 | $\cdots$ |
|---|---|---|---|---|---|---|---|---|---|---|
| $r$: | 4.0 | 3.5 | 2.2 | .30 | $-1.7$ | $-3.2$ | $-4.0$ | $-2.6$ | 1.1 | $\cdots$ |

The graph is shown in Figure 64.

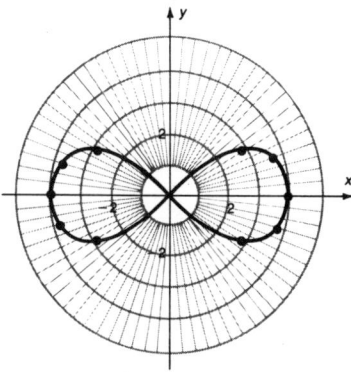

Figure 64  Graph of $r = 4 \cos 2\theta$  □

In general,

$$r = a \cos n\theta$$

is a **four-leaved rose** if $n = 2$ and the length of the leaves is $a$. If $n$ is an even number, the curve has $2n$ leaves; if $n$ is odd, the number of leaves is $n$. These leaves are equally spaced on a circle of radius $a$.

**EXAMPLE 9**

Graph $r = 4 \cos 2(\theta - \frac{\pi}{4})$.

*Solution*

This is a rose curve with four leaves of length 4 equally spaced on a circle. However, this curve has been rotated $\frac{\pi}{4}$, as shown in Figure 65.

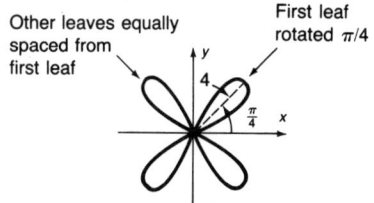

Figure 65  Graph of
$r = 4 \cos 2(\theta - \frac{\pi}{4})$  □

Notice that Example 9 can be rewritten as a sine curve:

$$r = 4 \cos 2(\theta - \tfrac{\pi}{4}) = 4 \cos(2\theta - \tfrac{\pi}{2}) = 4 \cos(\tfrac{\pi}{2} - 2\theta) = 4 \sin 2\theta$$

These steps can be reversed to graph a rose curve written in terms of a sine function.

**EXAMPLE 10**

*Solution*

Graph $r = 5 \sin 4\theta$.

$r = 5 \sin 4\theta = 5 \cos(\frac{\pi}{2} - 4\theta) = 5 \cos(4\theta - \frac{\pi}{2}) = 5 \cos 4(\theta - \frac{\pi}{8})$

Recognize this as a rose curve rotated $\frac{\pi}{8}$. There are eight leaves of length 5. The leaves are a distance of $\frac{\pi}{4}$ apart, as shown in Figure 66.

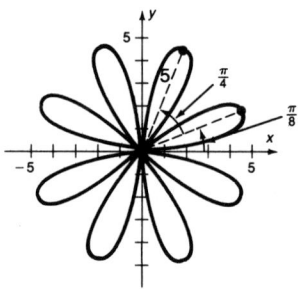

Figure 66   Graph of $r = 5 \sin 4\theta$    □

## 8.8 Lemniscates

The third (and last) general type of polar-form curve we will consider is called a **lemniscate**. It has the general form

$$r^2 = a^2 \cos 2\theta$$

**EXAMPLE 11**

*Solution*

Graph $r^2 = 16 \cos 2\theta$.

As before, when graphing a curve for the first time, begin by plotting points. For this example, be sure to obtain two values for $r$ when solving this quadratic equation. For example, if $\theta = 0$, then

$$\cos 2\theta = 1 \quad \text{and} \quad r^2 = 16, \quad \text{so} \quad r = 4 \quad \text{or} \quad -4$$

Also notice that for $\frac{\pi}{4} < \theta < \frac{3\pi}{4}$ there are no values for $r$, since $\cos 2\theta$ is negative. For $\pi \leq \theta < 2\pi$, the values repeat the sequence of values for $0 \leq \theta < \pi$. The points are connected as shown in Figure 67.

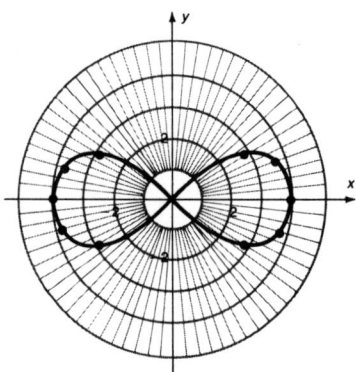

Figure 67   Graph of $r^2 = 16 \cos 2\theta$    □

The graph of $r^2 = a^2 \sin 2\theta$ is also a lemniscate. There are always two leaves to a lemniscate, and the length of the leaves is $a$. The sine function can be considered as a rotation of the cosine function.

**EXAMPLE 12**

Graph $r^2 = 16 \sin 2\theta$.

*Solution*

$$r^2 = 16 \sin 2\theta = 16 \cos(\tfrac{\pi}{2} - 2\theta) = 16 \cos(2\theta - \tfrac{\pi}{2}) = 16 \cos 2(\theta - \tfrac{\pi}{4})$$

This is a lemniscate, whose leaf has length $\sqrt{16} = 4$ and is rotated $\tfrac{\pi}{4}$, as shown in Figure 68.

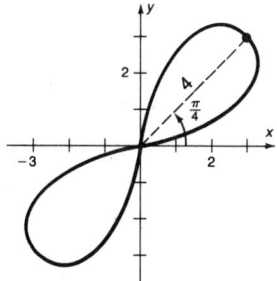

Figure 68  Graph of $r^2 = 16 \sin 2\theta$     □

### 8.9 Catalog of Polar-form Curves

We conclude this chapter by summarizing in Table 2 the special types of polar-form curves we have developed. There are many others, some of which are presented in the problems, but the three special types in Table 2 are by far the most common.

## Table 2   CATALOG OF POLAR-FORM CURVES

| CARDIOIDS | ROSE CURVES | LEMNISCATES |
|---|---|---|
| $r = a(1 \pm \cos\theta)$ <br> $r = a(1 \pm \sin\theta)$ <br><br> Notice that the points you plot are a distance of $a$ and $2a$ from the pole. | $r = a \cos n\theta$ <br> $r = a \sin n\theta$ <br><br> Notice that the length of each leaf is $a$. | $r^2 = a^2 \cos 2\theta$ <br> $r^2 = a^2 \sin 2\theta$ <br><br> Notice that the length of each leaf is $\sqrt{a^2} = a$. |

### CARDIOIDS

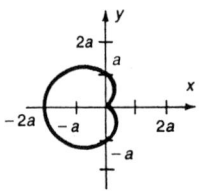

$r = a - a\cos\theta$
(no rotation)

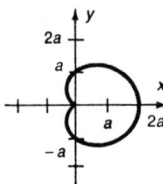

$r = a + a\cos\theta$
(180° rotation)

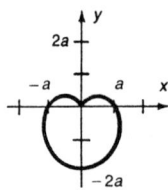

$r = a - a\sin\theta$
(90° rotation)

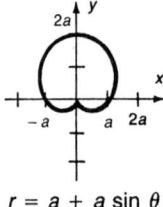

$r = a + a\sin\theta$
(270° rotation)

### ROSE CURVES

**1.  If $n$ is odd, the rose is $n$-leaved.**
*One leaf:* If $n = 1$, the rose is a curve with a single, circular leaf.

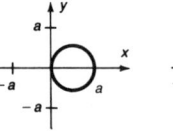

$r = a\cos\theta$
(no rotation)

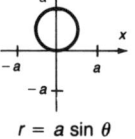

$r = a\sin\theta$
(90° rotation)

*Three leaves:*

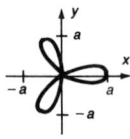

$r = a\cos 3\theta$     $r = a\sin 3\theta$
(no rotation)     (30° rotation)

**2.  If $n$ is even, the rose is $2n$-leaved.**
*Two leaves:* See the lemniscate.
*Four leaves:* $n = 2$

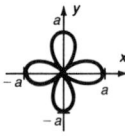
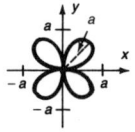

$r = a\cos 2\theta$     $r = a\sin 2\theta$
(no rotation)     (45° rotation)

### LEMNISCATES

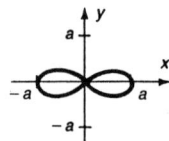

$r^2 = a^2 \cos 2\theta$
(no rotation)

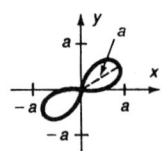

$r^2 = a^2 \sin 2\theta$
(45° rotation)

## 8.10  PROBLEM SET 8

**A**

*In Problems 1–10, plot each of the given polar-form points. Give both primary representations, and give the rectangular coordinates of the points. In Problems 7–10, approximate values to two decimal places.*

**1.** $(4, \frac{\pi}{4})$  **2.** $(6, \frac{\pi}{3})$  **3.** $(5, \frac{2\pi}{3})$  **4.** $(3, -\frac{\pi}{6})$  **5.** $(\frac{3}{2}, -\frac{5\pi}{6})$

**6.** $(5, -\frac{\pi}{2})$  **7.** $(-4, 4)$  **8.** $(4, 10)$  **9.** $(-4, 5\pi)$  **10.** $(-3, -\frac{7\pi}{3})$

*In Problems 11–20, plot the given rectangular-form points and give both primary representations in polar form. In Problems 17–20, approximate values to the nearest hundredth.*

**11.** $(5, 5)$  **12.** $(-1, \sqrt{3})$  **13.** $(2, -2\sqrt{3})$  **14.** $(-2, -2)$

**15.** $(3, -3)$  **16.** $(-6, 6)$  **17.** $(-\sqrt{3}, 1)$  **18.** $(4, 3)$

**19.** $(-12, 5)$  **20.** $(3, 7)$

*Identify each of the curves in Problems 21–32 as a cardioid, a rose curve (state the number of leaves), a lemniscate, or none of the above.*

**21.** $r^2 = 9 \cos 2\theta$  **22.** $r = 2 \sin 2\theta$  **23.** $r = 3 \sin 3\theta$

**24.** $r^2 = 2 \cos 2\theta$  **25.** $r = 2 - 2 \cos \theta$  **26.** $r = 3 + 3 \sin \theta$

**27.** $r^2 = \sin 3\theta$  **28.** $r = 4 \sin 30°$  **29.** $r = 5 \cos 60°$

**30.** $\theta = \tan \frac{\pi}{4}$  **31.** $r = 5(1 - \sin \theta)$  **32.** $\cos \theta = 1 - r$

*Sketch each of the curves whose equations are given in Problems 33–46. If it is not one of the types discussed in this chapter, graph it by plotting points.*

**33.** $r = 2(1 + \cos \theta)$  **34.** $r = 3(1 - \sin \theta)$

**35.** $r = 4(1 + \sin \theta)$  **36.** $r = 4 \cos 2\theta$

**37.** $r = 5 \sin 3\theta$  **38.** $r = 3 \cos 3\theta$

**39.** $r = 2 \cos \theta$  **40.** $r^2 = 9 \cos 2\theta$

**41.** $r^2 = 16 \cos 2\theta$  **42.** $r^2 = 16 \sin 2\theta$

**43.** $r = 5 \sin \frac{\pi}{6}$  **44.** $r = 9 \cos \frac{\pi}{9}$

**45.** $r = \theta$  **46.** $r = 3\theta$

**B**  **47.**  Derive the equations for changing from polar coordinates to rectangular coordinates.

**48.**  Derive the equations for changing from rectangular coordinates to polar coordinates.

**49.**  The **limaçon** is a curve of the form
$r = b \pm a \cos \theta$  or  $r = b \pm a \sin \theta$  where  $a > 0, b > 0$
There are four types of limaçons.
  **a.**  $b/a < 1$
      Graph $r = 2 - 3 \cos \theta$ by plotting points.
  **b.**  $b/a = 1$
      Graph $r = 2 - 2 \cos \theta$ by plotting points.

    **c.**    $1 < b/a < 2$
        Graph $r = 3 - 2 \cos \theta$ by plotting points.

    **d.**    $b/a \geq a$
        Graph $r = 3 - \cos \theta$ by plotting points.

**50.**    **Spirals** are interesting mathematical curves.   There are three general types of spirals.

    **a.**    *Spiral of Archimedes* has the form $r = a\theta$

        Graph $r = 2\theta$ for $\theta > 0$ by plotting points.

    **b.**    *Hyperbolic spiral* has the form $r\theta = a$

        Graph $r\theta = 2$ for $\theta > 0$ by plotting points.

    **c.**    *Logarithmic spiral* has the form $r = a^{k\theta}$

        Graph $r = 2^{\theta}$ by plotting points.

**51.**    The **stephoid** is a curve of the form $r = a \cos 2\theta \sec \theta$.   Graph this curve where $a = 2$, by plotting points.

**52.**    The **bifolium** has the form $r = a \sin \theta \cos^2 \theta$; graph this curve where $a = 1$, by plotting points.

**53.**    The **folium of Descartes** has the form

$$r = \frac{3a \sin \theta \cos \theta}{\sin^3 \theta + \cos^3 \theta}$$

Graph this curve where $a = 2$, by plotting points.

# PARAMETRIC EQUATIONS

**CHAPTER 9**

## 9.1 Definitions

Up to now, the curves we have discussed have been represented by a single equation. However, another way of representing curves is often useful. This new representation defines the $x$ and $y$ in $(x, y)$ so that they are *each* a function of some other variable, say $t$. For example, let

$$x = 1 + 3t, \; y = 2t \quad \text{for} \quad 0 \leq t \leq 5$$

Then, if $t = 1$,

$$x = 1 + 3(1) = 4 \quad \text{and} \quad y = 2(1) = 2$$

This point $(4, 2)$ can be plotted in the usual fashion; and we say that this point corresponds to the value $t = 1$. Other values are shown in the following table and are plotted in Figure 69.

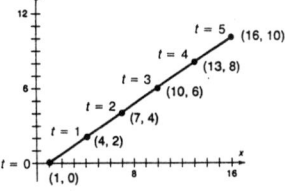

Figure 69 Graph of $x = 1 + 3t, \; y = 2t$

| $t$: | 0 | 1 | 2 | 3 | 4 | 5 |
|------|---|---|---|----|----|----|
| $x$: | 1 | 4 | 7 | 10 | 13 | 16 |
| $y$: | 0 | 2 | 4 | 6 | 8 | 10 |

The variable $t$ is called a **parameter**, and the equations $x = 1 + 3t$ and $y = 2t$ are called **parametric equations** for the line segment shown in Figure 69.

**PARAMETERS AND PARAMETRIC EQUATIONS**

Let $t$ be a number in an interval $I$. Consider the curve defined by the set of ordered pairs $(x, y)$, where
$$x = f(t) \quad \text{and} \quad y = g(t)$$
for functions $f$ and $g$ defined on $I$. Then the variable $t$ is called a **parameter**, and the equations $x = f(t)$ and $y = g(t)$ are called **parametric equations** for the curve defined by $(x, y)$.

**EXAMPLE 1**

Plot the curve represented by the parametric equations

$$x = \cos \theta, \quad y = \sin \theta$$

*Solution*

The parameter is $\theta$, and you can generate a table of values:

---
*Optional chapter

85

| $\theta$: | 0° | 15° | 30° | 45° | 60° | 75° | 90° | 120° | ⋯ |
|---|---|---|---|---|---|---|---|---|---|
| $x$: | 1.00 | .97 | .87 | .71 | .50 | .26 | .00 | − .50 | ⋯ |
| $y$: | .00 | .26 | .50 | .71 | .87 | .97 | 1.00 | .87 | ⋯ |

These points are plotted in Figure 70. If the plotted points are connected, you can see that the curve is a circle.

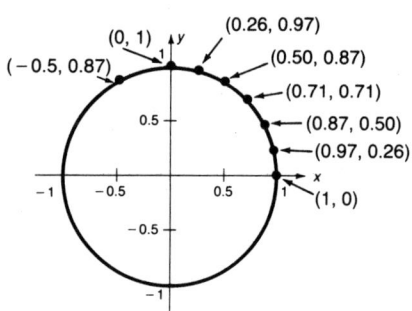

**Figure 70**   Graph of $x = \cos \theta$, $y = \sin \theta$   ❑

## 9.2  Eliminating the Parameter

It is possible to recognize the parametric equations in Example 1 as a unit circle, if you square both sides of the equation and add

$$x^2 = \cos^2 \theta \qquad y^2 = \sin^2 \theta$$
$$x^2 + y^2 = \cos^2 \theta + \sin^2 \theta = 1$$

The resulting equation is $x^2 + y^2 = 1$. This process is called **eliminating the parameter.**

**EXAMPLE 2**   Eliminate the parameter for the parametric equations $x = t + 2$, $y = t^2 + 2t - 1$.

*Solution*   Solve the first equation for $t$:   $t = x - 2$. Substitute this into the second (other) equation.

$$\begin{aligned} y &= (x - 2)^2 + 2(x - 2) - 1 \\ &= x^2 - 4x + 4 + 2x - 4 - 1 \\ &= x^2 - 2x - 1 \end{aligned}$$

You should recognize $y = x^2 - 2x - 1$ as a parabola. It can now be graphed by completing the square or by using the parametric equations and plotting points, as illustrated in Example 1.   ❑

**EXAMPLE 3**   Eliminate the parameter for the parametric equations

$$x = t^2 - 3t + 1, \ \ y = -t^2 + 2t + 3$$

*Solution*   It is not as easy to solve one of these equations for $t$ as it was in Example 2. You can, however, add one equation to the other:

$$x + y = -t + 4 \quad \text{or} \quad t = 4 - x - y$$

This can be substituted into either equation to give

$$x = (4 - x - y)^2 - 3(4 - x - y) + 1$$
$$= 16 - 4x - 4y - 4x + x^2 + xy - 4y + xy^2 - 12 + 3x + 3y + 1$$
$$= x^2 + 2xy + y^2 - 5x - 5y + 5$$

The curve whose equation is $x^2 + 2xy - 6x - 5y + 5 = 0$ is a rotated parabola, since $B^2 - 4AC = 4 - 4(1)(1) = 0$.  □

## 9.3  PROBLEM SET 9

**A**

*Plot the curves in Problems 1–16 by plotting points.*

**1.** $x = 4t$, $y = -2t$          **2.** $x = t + 1$, $y = 2t$

**3.** $x = t$, $y = 2 + \frac{2}{3}(t - 1)$          **4.** $x = t$, $y = 3 - \frac{3}{5}(t + 2)$

**5.** $x = 2t$, $y = t^2 + t + 1$          **6.** $x = 3t$, $y = t^2 - t + 6$

**7.** $x = t$, $y = t^2 + 2t + 3$          **8.** $x = t$, $y = 2t^2 - 5t + 6$

**9.** $x = 3\cos\theta$, $y = 3\sin\theta$          **10.** $x = 2\cos\theta$, $y = 2\sin\theta$

**11.** $x = 4\cos\theta$, $y = 3\sin\theta$          **12.** $x = 5\cos\theta$, $y = 2\sin\theta$

**13.** $x = t^2 + 2t + 3$, $y = t^2 + t - 4$

**14.** $x = t^2 - 2t + 3$, $y = t^2 - t + 4$

**15.** $x = t^2 + 3t - 4$, $y = 2t^2 + 4t - 1$

**16.** $x = 2t^2 + t + 6$, $y = t^2 + t + 6$

*Eliminate the parameters in Problems 17–32, and plot the resulting equations.*

**17.** $x = 4t$, $y = -2t$          **18.** $x = t + 1$, $y = 2t$

**19.** $x = t$, $y = 2 + \frac{2}{3}(t - 1)$          **20.** $x = t$, $y = 3 - \frac{3}{5}(t + 2)$

**21.** $x = 2t$, $y = t^2 + t + 1$          **22.** $x = 3t$, $y = t^2 - t + 6$

**23.** $x = t$, $y = t^2 + 2t + 3$          **24.** $x = t$, $y = 2t^2 - 5t + 6$

**25.** $x = 3\cos\theta$, $y = 3\sin\theta$          **26.** $x = 2\cos\theta$, $y = 2\sin\theta$

**27.** $x = 4\cos\theta$, $y = 3\sin\theta$          **28.** $x = 5\cos\theta$, $y = 2\sin\theta$

**29.** $x = t^2 + 2t + 3$, $y = t^2 + t - 4$

**30.** $x = t^2 - 2t + 3$, $y = t^2 - t + 4$

**31.** $x = t^2 + 3t - 4$, $y = 2t^2 + 4t - 1$

**32.** $x = 2t^2 + t + 6$, $y = t^2 + t + 6$

*Plot the curves in Problems 33–40 by any convenient method.*

**33.** $x = 60t$, $y = 80t - 16t^2$          **34.** $x = 30t$, $y = 60t - 9t^2$

**35.** $x = 10\cos t$, $y = 10\sin t$          **36.** $x = 8\sin t$, $y = 8\cos t$

**37.** $x = 5\cos\theta$, $y = 3\sin\theta$          **38.** $x = 4\cos\theta$, $y = 2\sin\theta$

**39.** $x = t^2$, $y = t^3$          **40.** $x = t^3 + 1$, $y = t^3 - 1$

**B**  **41.**    Suppose a light is attached to the edge of a bike wheel.  The path of the light is shown in Figure 71.

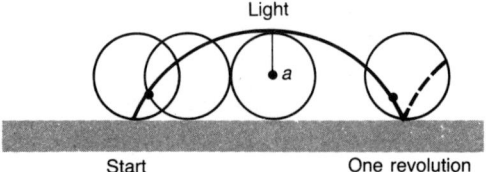

Figure 71  Graph of a cycloid

If the radius of the wheel is $a$, find the equation for the path of the light. Such a curve is called a **cycloid**.

*Hint*: Consider Figure 71. and find the coordinates of $P(x, y)$. Notice that

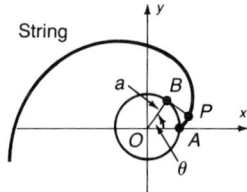

$$x = |OA|$$

$$y = |PA|$$

Find $x$ and $y$ in terms of $\theta$, the amount of rotation in radians.

**42.**    Suppose a string is wound around a circle of radius $a$. The string is then unwound in the plane of the circle while it is held tight, as shown in Figure 72.

Figure 72  Graph of the involute of a circle

Find the equation for this curve, called the **involute of a circle.**

*Hint*: Consider Figure 72 and find the coordinates of $P(x, y)$. Notice that

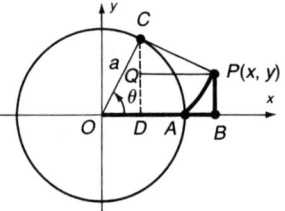

$$x = |OB|$$

$$y = |PB|$$

Find $x$ and $y$ in terms of $\theta$, the amount of rotation in radians.

# CHAPTER 10

# SUMMARY AND REVIEW

In analytic geometry we are concerned with relationships between algebra and geometry. We relate sets of ordered pairs satisfying an equation with the set of ordered pairs on a curve:

> By the *graph of an equation* or the *equation of a graph*, we mean that there is a one-to-one correspondence between ordered pairs satisfying the equation and ordered pairs lying on the graph.

There are two main problems for consideration:

1. Given an equation, find the graph.
2. Given a graph, find the corresponding equation.

## 10.1 Objectives

I. Lines
   - A. Define and find the slope of a line.
   - B. Find the length of a line segment.
   - C. Identify parallel and perpendicular lines.
   - D. Know the forms for equations of lines.
     1. Standard form: $Ax + By + C = 0$
     2. Slope–intercept form: $y = mx + b$
     3. Point–slope form: $y - y_1 = m(x - x_1)$
     4. Two-point form: $y - y_1 = \left(\dfrac{y_2 - y_1}{x_2 - x_1}\right)(x - x_1)$
   - E. Problem 1: Given an equation, graph the line.
   - F. Problem 2: Given the line, find the equation.

II. Conic Sections
   - A. Define parabola, ellipse, circle, and hyperbola.
   - B. Be familiar with the associated terminology: focus, directrix, center, axis, vertex, focal chord, asymptote, major axis, minor axis, transverse axis, conjugate axis.
   - C. Identify the conic by looking at the equation.
   - D. Know the standard forms of the equations of a conic.
     1. General form:
        $$Ax^2 + Bxy + Cy^2 + Dx + Ey + F = 0$$

2.    Standard-form parabolas:
opens up:    $(x - h)^2 = 4c(y - k)$
opens down:  $(x - h)^2 = -4c(y - k)$
opens right: $(y - k)^2 = 4c(x - h)$
opens left:  $(y - k)^2 = -4c(x - h)$

3.    Standard-form ellipses:

horizontal axis:    $\dfrac{(x - h)^2}{a^2} + \dfrac{(y - k)^2}{b^2} = 1$

vertical axis:    $\dfrac{(y - k)^2}{a^2} + \dfrac{(x - h)^2}{b^2} = 1$

4.    Standard-form hyperbolas:

horizontal axis:    $\dfrac{(x - h)^2}{a^2} - \dfrac{(y - k)^2}{b^2} = 1$

vertical axis:    $\dfrac{(y - k)^2}{a^2} - \dfrac{(x - h)^2}{b^2} = 1$

5.    Standard-form circle: $(x - h)^2 + (y - k)^2 = r^2$

E.    Problem 1: Given an equation, graph the conic.
F.    Problem 2: Given the curve, find the equation.
G.    The rotation $\theta$ for the general-form equation is given by

$$\cot 2\theta = \frac{A - C}{B}, \quad \text{where} \quad B \neq 0$$

H.    The equations for rotating the axes are
$x = x' \cos \theta - y' \sin \theta$
$y = x' \sin \theta + y' \cos \theta$

III.    **Curve Sketching:  Given an Equation, Sketch the Graph**
A.    Symmetry
1.    With respect to the $x$-axis; the equation is unchanged when $-y$ is substituted for $y$.
2.    With respect to the $y$-axis; the equation is unchanged when $-x$ is substituted for $x$.
3.    With respect to the origin; the equation is unchanged when $(-x, -y)$ is substituted for $(x, y)$.
B.    Extent
1.    Domain:   solve for $y$; it is the set of all real numbers except values that cause division by zero or a negative under a square root (or an even-indexed root).
2.    Range:   solve for $x$; it is the set of all real numbers except values that cause division by zero or a negative under a square root (or an even-indexed root).
C.    Asymptotes
1.    Horizontal:  solve for $x$; values $y = a$ where $a$ is a value that causes division by 0.
2.    Vertical:  solve for $y$; values $x = b$, where $b$ is a value that causes division by 0.
D.    Intercepts
1.    $x$-intercepts:  set $y = 0$ and solve.
2.    $y$-intercepts:  set $x = 0$ and solve.

E.    Plot points.

*IV.    Polar Curves

    A.    Plot polar-form points.

    B.    Primary representations are $(r, \theta)$ where $0 \leq \theta < 2\pi$ and $(-r, \theta + \pi)$ where $0 < \pi + \theta < 2\pi$.

    C.    Changing forms:

        1.    Polar to rectangular: $x = r \cos \theta$ and $y = r \sin \theta$

        2.    Rectangular to polar:

$$r = \sqrt{x^2 + y^2} \text{ and } \theta' = \tan^{-1}\left|\frac{y}{x}\right|$$

        where $x \neq 0$ and $\theta'$ is the reference angle for $\theta$.

    D.    Polar curves

        1.    Cardioid: $r = a(1 \pm \cos \theta)$ or $r = a(1 \pm \sin \theta)$

        2.    Rose curve: $r = a \cos n\theta$ or $r = a \sin n\theta$

            a.    If $n$ is odd, the rose is $n$-leaved

            b.    If $n$ is even, the rose is $2n$-leaved

        3.    Lemniscate: $r^2 = a^2 \cos 2\theta$ or $r^2 = a^2 \sin 2\theta$

*V.    Parametric Equations

    A.    Graph parametric equations by plotting points.

    B.    Graph parametric equations by eliminating the parameter.

## 10.2  Sample Test I

*In Problems 1–4, identify the curve.*

1.    **a.**  $xy + x^2 - 3x = 5$

    **b.**  $x^2 + y^2 + xy + 3x - y = 3$

    **c.**  $3x - 2y^2 - 4y + 7 = 0$

    **d.**  $\frac{x}{16} + \frac{y}{4} = 1$

2.    **a.**  $x^2 + 2xy + y^2 = 10$

    **b.**  $25x^2 + 16y^2 = 400$

    **c.**  $(x - 1)(y + 1) = 7$

    **d.**  $3x + 4y - 6 = 0$

*3.    **a.**  $x = 2 + 5t, y = -1 - 3t$

    **b.**  $2x^2 - y^2 + 4xy - 2x + 3y = 6$

    **c.**  $r = 2 + 2 \sin \theta$

    **d.**  $r^2 = 2 \cos 2\theta$

*4.    **a.**  $r = 4 \cos \frac{\pi}{4}$

    **b.**  $\theta = \tan \frac{\pi}{3}$

    **c.**  $r = 3 \cos 4\theta$

    **d.**  $r = 5 \sin \theta$

---

*Optional chapters

*Write the equations of the curves in Problems 5–9.*

5.  The ellipse with center at (4, 1), a focus at (5, 1), and a semimajor axis of 2.

6.  The line passing through (4, 7) and $(-1, -3)$.

7.  The parabola whose vertex is (6, 3) and with directrix $x = 1$.

8.  The hyperbola with center at $(-5, 4)$, a focus at (0, 4), and eccentricity 2.

9.  The set of points the difference of whose distances from $(-3, 4)$ and $(-7, 4)$ is 2.

*Graph the curves in Problems 10–15.*

10.  $25x^2 + 16y^2 = 400$        11.  $x^2 - y^2 + x - y = 3$

12.  $x^2 = y$                     13.  $x^2 + y^2 = 4x + 2y - 3$

14.  $x(x - y) = y(y - x) - 1$     *15.  $r - 4 = 4 \cos \theta$

16.  Find the equation of all points whose distances from the point (0, 8) are equal to their distances from the $x$-axis.

17.  a.  What are the focus, center, and directrix for the parabola $(y - 2)^2 = 24(x + 3)$?

    b.  Find the center, foci, eccentricity, length of major axis, and length of minor axis for the ellipse
    $$5x^2 + 4y^2 - 60x + 8y + 164 = 0$$

18.  a.  Find the angle of rotation for
    $$5x^2 + 4xy + 5y^2 + 3x - 2y + 5 = 0$$

    b.  Find the tangent of the angle of rotation for
    $$4x^2 + 4xy + y^2 + 3x - 2y + 7 = 0$$

19.  Given $2x + xy^2 - y = 0$:
    a.  What is the domain?
    b.  What is the range?

20.  For the curve given in Problem 19:
    a.  What is the symmetry for the given curve?
    b.  What are the horizontal or vertical asymptotes?

## 10.3  Sample Test 2

*In Problems 1–6, identify each curve.*

1.  a.  $\dfrac{x - 3}{2} = \dfrac{y + 5}{4}$       b.  $\dfrac{x - 5}{-1} = \dfrac{(y - 1)^2}{3}$

2.  a.  $2x^2 - 3xy + y^2 + 3x = 0$       b.  $2x - 3y + 3 = 0$

3.  a.  $y^2 = x^2 - 1$       b.  $x^2 = 4y^2 + 3x - 2y + 3$

4.  a.  $x^2 + y^2 + x - y = 3$       b.  $\dfrac{(x - 2)^2}{4} - \dfrac{(y + 1)}{9} = 1$

5.  a.  $x^2 + y + 2x + 1 = 0$       b.  $2x^2 + 4xy + 2y^2 + 3y = 0$

*6.  a.  $\cos \theta = \dfrac{r - 4}{4}$       b.  $r^2 = 4 \sin 2\theta$

---

*Optional chapters

*Write the equations of the curves in Problems 7–10.*

7. The set of points the sum of whose distances from $(-5, -2)$ and $(-3, -2)$ is 8.
8. The set of points equally distant from the point $(-3, 4)$ and the line $3x + 4y - 3 = 0$.
9. Find the equations of the line passing thorough $P(4, 6)$ and $Q(7, -1)$.
10. Eliminate the parameter: $x = 4 - 3t$, $y = -3 + t$.

*Graph the curves whose equations are given in Problems 11–15.*

11. $9x^2 + 25y^2 = 225$

12. $x^2 + y^2 + 2x - 2y = 3$

13. $x^2 - y + 6x + 2 = 0$

14. $x^2 - 3y^2 + 4x + 12y - 11 = 0$

15. $(4 - x)(x - y) = y(x + y)$

16. Find the equation of the line passing through $(1, -2)$ and $(-2, 3)$.
17. Given $x^2 + 4xy + 5y^2 = 9$:
    a. What is the appropriate rotation?
    b. What are the intercepts?
18. Given $x + x^2y + y = 0$:
    a. What is the domain?
    b. What is the range?
19. For the curve given in Problem 18:
    a. What is the symmetry?
    b. What are the horizontal or vertical asymptotes?
20. a. Find the center, foci, eccentricity, length of major axis, and length of minor axis for the ellipse

    $$3(x - 2)^2 + 4(y + 1)^2 = 12$$

    b. Find the center, foci, eccentricity, length of transverse axis, and length of conjugate axis for the hyperbola

    $$25x^2 - 9y^2 - 250x - 72y + 256 = 0$$

# APPENDIX A

# SOLVING EQUATIONS

## A.1 Linear Equations

To solve the first-degree (or linear) equation, isolate the variable on one side. That is, the solution of $ax + b = 0$ is

$$x = -\frac{b}{a}, \quad \text{where} \quad a \neq 0$$

To solve a linear equation, use the following steps.

**Step 1:** Use the distributive property to clear the equation of parentheses. If it is a rational expression, multiply both sides by the appropriate expression to eliminate the denominator in the problem. Be sure to check the solutions obtained in the original.

**Step 2:** Add the same number to both sides of the equality to obtain an equation in which all of the terms involving the variable are on one side and all of the other terms are on the other side.

**Step 3:** Multiply (or divide) both sides of the equation by the same nonzero number to isolate the variable on one side.

**EXAMPLE 1**

Solve $4(x - 3) + 5x = 5(8 + x)$.

*Solution*

$$
\begin{aligned}
4(x - 3) + 5x &= 5(8 + x) \\
4x - 12 + 5x &= 40 + 5x \\
9x - 12 &= 40 + 5x \\
4x - 12 &= 40 \\
4x &= 52 \\
x &= 13
\end{aligned}
$$
$\square$

**EXAMPLE 2**

Solve $\dfrac{x + 1}{x - 2} = \dfrac{x + 2}{x - 2}$.

*Solution*

$$
\begin{aligned}
(x + 1)(x - 2) &= (x + 2)(x - 2) \\
x^2 - 2x + x - 2 &= x^2 - 4 \\
-x - 2 &= -4 \\
-x &= -2 \\
x &= 2
\end{aligned}
$$

Notice that $x = 2$ causes division by 0, so the solution set is empty. $\square$

**EXAMPLE 3**

Solve $4x + 5xy + 3y^2 = 10$ for $x$.

*Solution*

$$4x + 5xy = 10 - 3y^2$$

$$(4 + 5y)x = 10 - 3y^2$$

$$x = \frac{10 - 3y^2}{4 + 5y}, \quad y \neq -\frac{4}{5} \qquad \square$$

### A.2 Quadratic Equations

To solve a second-degree (or quadratic) equation, first obtain a zero on one side. Next, try to factor the quadratic. If is factorable, set each factor equal to zero and solve. If it is not factorable, use the quadratic formula:

**The solution of $ax^2 + bx + c = 0$ is**

$$x = \frac{-b \pm \sqrt{b^2 - 4ac}}{2a} \quad \text{where} \quad a \neq 0$$

**EXAMPLE 4**

Solve $2x^2 - x - 3 = 0$.

*Solution*

Factoring, $(2x - 3)(x + 1) = 0$. Since $2x - 3 = 0$, $x = \frac{3}{2}$; since $x + 1 = 0$, $x = -1$. The solution is $x = \frac{3}{2}, -1$. $\qquad \square$

**EXAMPLE 5**

Solve $5x^2 - 3x - 4 = 0$.

*Solution*

$a = 5$, $b = -3$, and $c = -4$; thus (since it does not factor),

$$x = \frac{3 \pm \sqrt{9 - 4(5)(-4)}}{2(5)}$$

$$= \frac{3 \pm \sqrt{89}}{10} \qquad \square$$

**EXAMPLE 6**

Solve $x^2 + 2xy + 3y^2 - 4 = 0$.

*Solution*

$a = 1$, $b = 2y$, $c = 3y^2 - 4$; thus,

$$x = \frac{-2y \pm \sqrt{4y^2 - 4(1)(3y^2 - 4)}}{2(1)}$$

$$= \frac{-2y \pm \sqrt{4y^2 - 12y^2 + 16}}{2}$$

$$= \frac{-2y \pm \sqrt{16 - 8y^2}}{2}$$

$$= -y \pm \sqrt{4 - 2y^2} \qquad \square$$

# APPENDIX B     SOLVING INEQUALITIES

### B.1  Linear Inequalities

The first-degree inequality is solved following the steps outlined in Appendix A for solving first-degree equations.  The only difference is that, when we multiply or divide both sides of an inequality by a negative value, the order of the inequality is reversed.  That is, if $a < b$, then

$$a + c < b + c \qquad \text{for any number } c$$
$$a - c < b - c \qquad \text{for any number } c$$
$$ac < bc \qquad \text{for any positive number } c$$
$$ac > bc \qquad \text{for any negative number } c$$

A similar result holds if we use $\leq$, $>$, or $\geq$.

Answers to inequality problems are often intervals on a real number line.  We use the following notation for intervals:

*closed interval* (endpoints included): $[a, b]$

*open interval* (endpoints excluded): $(a, b)$

$(a, \infty)$

*half-open (or half-closed) interval:* $(a, b]$

$[a, b)$

$(-\infty, a]$

To denote an interval that is not connected, we use the notation for the union of two sets:

$[a, b) \cup (c, d]$

$[a, b) \cup (b, c]$

**EXAMPLE 1**

*Solution*

Solve $5x + 3 < 3x - 15$.

$$2x + 3 < -15$$
$$2x < -18$$
$$x < -9$$

The solution is $(-\infty, -9)$.                                    □

**EXAMPLE 2**        Solve $(2x + 1)(x - 5) \le (2x - 3)(x + 2)$.

*Solution*              $2x^2 - 9x - 5 \le 2x^2 + x - 6$

                            $-9x - 5 \le x - 6$

                            $-10x \le -6$

                            $x \ge \frac{1}{10}$

The solution set is $[.1, \infty)$.                                      □

**EXAMPLE 3**        Solve $-5 \le x + 4 \le 5$.

*Solution*              $-9 \le x \le 1$

The solution set is $[-9, 1]$.                                      □

**B.2  Quadratic Inequalities**

The method of solving quadratic inequalities is similar to the method of solving quadratic equalities.

**Step 1:** Obtain a zero on one side of the inequality.

**Step 2:** Factor, if possible. If the inequality is a rational expression, factor the numerator and the denominator separately.

**Step 3:** Set each factor equal to zero. These values are not necessarily the solution of the inequality. If the inequality is not factorable, we treat the entire expression as a single factor and solve by the quadratic formula. Values for which the factors are zero are called the *critical values of x*. Plot these values on a number line. The points determine one or more intervals on a number line.

**Step 4:** Choose some value in each interval. It will make the inequality true or false; we accordingly include or exclude that interval from the solution set.

**EXAMPLE 4**        Solve $x^2 < 6 - x$.

*Solution*              $x^2 + x - 6 < 0$
                            $(x - 2)(x + 3) < 0$

Factors            Signs
$x - 2$:         $- -$ | $- -$ | $+ +$
$x + 3$:         $- -$ | $+ +$ | $+ +$
                  *pos*  |  *neg*  |  *pos*    ← *signs of product*

critical values → $-3$     $2$

We see from the number line that the solution is $(-3, 2)$. The endpoints are not included, so the solution is an open interval. These steps are summarized as:

$- -$    $- +$    $+ +$       ← *signs of factors*

*a*        *b*

*pos*    *neg*    *pos*       ← *sign of product (negative because of* <)

                                                                □

**EXAMPLE 5**        Solve $(2 - x)(x + 3)(x - 1) \geq 0$.

*Solution*        Plotting critical values and checking the signs of the factors, we have

$$+--\qquad ++-\qquad +++\quad -++$$

*Endpoints included because of $\geq$*

The solution is $(-\infty, -3] \cup [1, 3]$.        □

The method of solving quadratic inequalities also applies when we wish to solve inequalities involving rational expressions.

**EXAMPLE 6**        Solve $\dfrac{x + 3}{x - 2} < 0$.

*Solution*        Be careful not to multiply both sides by $(x - 2)$, since you do not know whether $(x - 2)$ is positive or negative. You could consider separate cases, but instead solve it as if it were a quadratic inequality. We set the numerator and denominator each equal to 0 to obtain the critical values $x = -3$, $x = 2$. We then plot these on a number line and check a value in each interval to determine the solution.

$$--\quad +-\quad ++$$

The solution is $(-3, 2)$.        □

**EXAMPLE 7**        Solve $\dfrac{x - 3}{x} > 1$.

*Solution*        
$$\frac{x - 3}{x} - 1 > 0$$
$$\frac{x - 3 - x}{x} > 0$$
$$\frac{-3}{x} > 0$$

The solution is $(-\infty, 0)$.        □

**EXAMPLE 8**        Solve $\dfrac{x + 2}{2x} \geq 5$.

*Solution*        
$$\frac{x + 2}{2x} - 5 \geq 0$$
$$\frac{x + 2 - 10x}{2x} \geq 0$$
$$\frac{2 - 9x}{2x} \geq 0$$

The solution is $(0, \frac{2}{9}]$. The endpoints are included when we have intervals with $\geq$ or $\leq$. However, values that cause division by zero are excluded.        □

**EXAMPLE 9**        Solve $x^2 + 2x - 4 < 0$

*Solution*        The left-hand expression is in simplified form and cannot be factored. Therefore, we proceed by considering $(x^2 + 2x - 4)$ as a single factor.

To find the critical values, we find the values for which the factor is 0.

$$x^2 + 2x - 4 = 0$$

$$x = \frac{-2 \pm \sqrt{4 - 4(1)(-4)}}{2}$$

$$= -1 \pm \sqrt{5}$$

Plot the critical values, and check the sign of the expression in each of the intervals.

The solution is $(-1 - \sqrt{5}, -1 + \sqrt{5})$.　□

# APPENDIX C   COMPLETING THE SQUARE

In algebra, the method of completing the square was first introduced as a technique for solving quadratic equations. This method leads to a proof of the quadratic formula (see Appendix A). However, when we are working with conic sections, we need to complete the square whenever the conic is not centered at the origin. Consider

$$Ax^2 + Bx + C = 0$$

We wish to put this into the form

$$(x + ?)^2 = \text{some number}$$

To do this, we follow the steps below:

**Step 1:** Subtract $C$ (the constant term) from both sides.

$$Ax^2 + Bx = -C$$

**Step 2:** Divide both sides by $A$ ($A \neq 0$; if $A = 0$, the equation would not be quadratic). That is, we want the coefficient of the squared term to be 1.

$$x^2 + \frac{B}{A}x = -\frac{C}{A}$$

**Step 3:** Add $\left(\frac{B}{2A}\right)^2$ to both sides. That is, take one-half of the coefficient of the first-degree term, square it, and add it to both sides.

$$x^2 + \frac{B}{A}x + \left(\frac{B}{2A}\right)^2 = \frac{B^2}{4A^2} - \frac{C}{A}$$

**Step 4:** The expression on the left is not a "perfect square" and can be factored.

$$\left(x + \frac{B}{2A}\right)^2 = \frac{B^2}{4A^2} - \frac{C}{A}$$

**EXAMPLE 1**   Complete the square for $x^2 + 2x - 5 = 0$.

*Solution*

$$x^2 + 2x - 5 = 0$$
$$x^2 + 2x = 5$$
$$x^2 + 2x + 1 = 5 + 1$$
$$(x + 1)^2 = 6$$

□

**EXAMPLE 2**

*Solution*

Complete the square for $3x^2 + 5x - 4 = 0$.

$$3x^2 + 5x - 4 = 0$$
$$3x^2 + 5x = 4$$
$$x^2 + \tfrac{5}{3}x = \tfrac{4}{3}$$
$$x^2 + \tfrac{5}{3}x + \left(\tfrac{5}{6}\right)^2 = \tfrac{4}{3} + \tfrac{25}{36}$$
$$\left(x + \tfrac{5}{6}\right)^2 = \tfrac{73}{36}$$

    □

**EXAMPLE 3**

Complete the square in both the $x$ and the $y$ terms.

$$x^2 + y^2 + 4x - 9y - 13 = 0$$

*Solution*

Associate the $x$ and $y$ terms:

$$(x^2 + 4x) + (y^2 - 9y) = 13$$
$$(x^2 + 4x + 4) + (y^2 - 9y + \tfrac{81}{4}) = 13 + 4 + \tfrac{81}{4}$$
$$(x + 2)^2 + (y - \tfrac{9}{2})^2 = \tfrac{149}{4}$$

    □

**EXAMPLE 4**

Complete the square in both the $x$ and the $y$ terms.

$$2x^2 - 3y^2 - 12x + 6y + 7 = 0$$

*Solution*

Associate the $x$ and $y$ terms:

$$(2x^2 - 12x) + (-3y^2 + 6y) = -7$$

In this example, we cannot divide by the coefficient of the squared terms, because we need to make *both* the $x^2$ and the $y^2$ coefficients equal to one. Instead, we factor

$$2(x^2 - 6x) - 3(y^2 - 2y) = -7$$

Now we complete the square; be sure you add the *same number* to both sides.

$$\underbrace{2(x^2 - 6x + 9)}_{\text{add 18 to both sides}} - \underbrace{3(y^2 - 2y + 1)}_{\text{add } -3 \text{ to both sides}} = -7 + 18 - 3$$

$$2(x - 3)^2 - 3(y - 1)^2 = 8$$

    □

# APPENDIX D    SELECTED ANSWERS

## Problem Set 1

1. **b.** $9\sqrt{2} + \sqrt{82}$  **c.** $d_{AB} = \sqrt{50}$; $d_{BC} = \sqrt{32}$; $d_{AC} = \sqrt{82}$. These distances satisfy the Pythagorean Theorem, so the triangle is a right triangle.  **d.** $m_{AB} = 1$ and $m_{BC} = -1$, so the segments are perpendicular. Thus, the triangle is a right triangle.  **2. b.** $8\sqrt{5}$  **c.** $m_{AB} = m_{CD} = \frac{1}{2}$; $m_{AD} = m_{BC} = -2$. Thus, the opposite sides of the quadrilateral are parallel, so the figure is a parallelogram.  **d.** The adjacent sides have slopes $\frac{1}{2}$ and $-2$ so they are perpendicular. Thus, it is a rectangle.  **3.** $4x + y - 14 = 0$  **4.** $3x + 4y - 23 = 0$  **5.** $y + 3 = 0$

6. $x + 2 = 0$  **7.** $x - 6y - 24 = 0$  **8.** $2x + 5y - 15 = 0$  **9.** $4x - 3y - 8 = 0$

10. $5x - 2y + 15 = 0$  **11.** $x + y - 5 = 0$  **12.** $4x - 3y - 17 = 0$  **13.** $4x - 5y + 20 = 0$

14. $5x + 2y + 10 = 0$  **15.** $x + 4y - 8 = 0$  **16.** $2x + y - 17 = 0$  **17.** $x - 4y = 0$

18. $2x - 3y - 11 = 0$  **19.** $x - 5y + 14 = 0$  **20.** $2x + 3y + 5 = 0$  **27.** $(0, -\frac{1}{6})$  **28.** $(\frac{1}{2}, 0)$

29. $(0, -\frac{C}{B})$ and $m = -\frac{A}{B}$ if $B \neq 0$; thus the line is vertical, with no slope and no $y$-intercept.

30. $x + y - 6 = 0$

## Problem Set 2

1. Opens right; center $(0, 0)$; length of focal chord, $4c = 8$; $c = 2$.  **2.** Opens left; center $(0, 0)$; length of focal chord, $4c = 12$; $c = 3$.  **3.** Opens left; center $(0, 0)$; length of focal chord, $4c = 20$; $c = 5$.  **4.** Opens upward; center $(0, 0)$; length of focal chord, $4c = \frac{5}{2}$; $c = \frac{5}{8}$.  **5.** Opens downward; center $(0, 0)$; length of focal chord, $4c = 4$; $c = 1$.  **6.** Opens downward; center $(0, 0)$; length of focal chord, $4c = 2$; $c = \frac{1}{2}$.  **7.** Opens downward; center $(0, 0)$; length of focal chord, $4c = \frac{5}{2}$; $c = \frac{5}{8}$.

8. Opens left; center $(0, 0)$; length of focal chord, $4c = 3$; $c = \frac{3}{4}$.  **9.** Opens right; center $(0, 0)$; length of focal chord, $4c = 5$; $c = \frac{5}{4}$.  **10.** Opens left; center $(-4, 0)$; length of focal chord, $4c = 3$; $c = \frac{3}{4}$.  **11.** Opens right; center $(-2, 1)$; length of focal chord, $4c = 2$; $c = \frac{1}{2}$.  **12.** Opens right; center $(1, -3)$; length of focal chord, $4c = 3$; $c = \frac{3}{4}$.  **13.** Opens upward; center $(-2, 1)$; length of focal chord, $4c = 2$; $c = \frac{1}{2}$.  **14.** Opens upward; center $(1, -3)$; length of focal chord, $4c = \frac{3}{4}$; $c = \frac{3}{16}$.

15. This is a line passing through $(1, -2)$ with slope $-\frac{1}{2}$.  **16.** This is a line passing through

$(-3, 1)$ with slope $-\frac{1}{3}$.    **17.** Opens right; center $(-3, -5)$; length of focal chord, $4c = 4$; $c = 1$.

**18.** Opens left; center $(-\frac{5}{16}, \frac{3}{2})$; length of focal chord, $4c = 4$; $c = 1$.    **19.** Opens right; center

$(7, -2)$; length of focal chord, $4c = 10$; $c = \frac{5}{2}$.    **20.** Opens downward; center $(3, -1)$; length of focal

chord, $4c = 9$; $c = \frac{9}{2}$.    **21.** Opens right; center $(\frac{7}{3}, -2)$; length of focal chord, $4c = \frac{3}{2}$; $c = \frac{3}{8}$.

**22.** Opens downward; center $(-\frac{1}{3}, \frac{13}{9})$; length of focal chord, $4c = 2$; $c = \frac{1}{2}$.    **23.** $y^2 = 10(x - \frac{5}{2})$

**24.** $x^2 = -6(y + \frac{3}{2})$    **25.** $(y - 2)^2 = -16(x + 1)$    **26.** $(x - 4)^2 = 12(y + 1)$

**27.** $(x + 2)^2 = 24(y + 3)$    **28.** $(y - 4)^2 = 16(x + 3)$    **29.** $x = 2$    **30.** $3x + y - 5 = 0$

**31.** $(x + 3)^2 = -\frac{1}{3}(y - 2)$    **32.** $(y - 2)^2 = -\frac{36}{7}(x - 4)$

**33.** $x^2 + 2xy + y^2 - 22x + 18y + 41 = 0$    **34.** $131x^2 - 120xy + 12y^2 + 174x - 170y - 428 = 0$

**35.** $(x - 100)^2 = -200(y - 50)$

## Problem Set 3

**1.** Circle; center $(0, 0)$; $r = 1$.    **2.** Vertical ellipse; center $(0, 0)$; $a = 5$; $b = 4$.    **3.** Vertical ellipse;

center $(0, 0)$; $a = 3$; $b = 2$.    **4.** Vertical ellipse; center $(0, 0)$; $a = 6$; $b = 5$.    **5.** Vertical ellipse;

center $(0, 0)$; $a = 6$; $b = 5$.    **6.** Vertical ellipse; center $(0, 0)$; $a = \sqrt{3}$; $b = \sqrt{2}$.    **7.** Vertical ellipse;

center $(0, 0)$; $a = 2$; $b = \sqrt{3}$.    **8.** Vertical ellipse; center $(0, 0)$; $a = \frac{1}{10}\sqrt{70}$; $b = \frac{1}{5}\sqrt{35}$.    **9.** Circle;

center $(2, -3)$; $r = 5$.    **10.** Horizontal ellipse; center $(-1, 1)$; $a = 2$; $b = \sqrt{3}$.    **11.** Horizontal

ellipse; center $(-3, 1)$; $a = 9$; $b = 3$.    **12.** Horizontal ellipse; center $(3, 2)$; $a = 4$; $b = 3$.

**13.** Circle; center $(-2, -3)$; $r = 5$.    **14.** Vertical ellipse; center $(1, -2)$; $a = 3$; $b = 2$.

**15.** Vertical ellipse; center $(-3, 2)$; $a = 4$; $b = 3$.    **16.** Horizontal ellipse; center $(-7, -3)$; $a = 1$;

$b = \frac{1}{5}$.    **17.** Horizontal ellipse; center $(-\frac{1}{3}, 1)$; $a = \frac{1}{3}$; $b = \frac{1}{6}\sqrt{3}$.    **18.** Vertical ellipse; center $(\frac{1}{4}, -\frac{1}{3})$;

$a = \frac{1}{3}\sqrt{3}$; $b = \frac{1}{6}\sqrt{6}$.    **19.** $(x - 4)^2 + (y - 5)^2 = 36$    **20.** $\frac{x^2}{100} + \frac{y^2}{64} = 1$    **21.** $\frac{x^2}{24} + \frac{y^2}{49} = 1$

**22.** $\frac{(x - 4)^2}{7} + \frac{(y + 1)^2}{16} = 1$    **23.** $(x + 3)^2 + (y - 2)^2 = 16$    **24.** $\frac{16(x + 1)^2}{225} + \frac{16(y + 3)^2}{81} = 1$

**25.** $(x - 2)^2 + (y + 3)^2 = 25$    **26.** $\frac{x^2}{89} + \frac{16y^2}{89} = 1$

## Problem Set 4

**1.** Horizontal hyperbola; center $(0, 0)$; $a = 1$; $b = 1$.    **2.** Horizontal hyperbola; center $(0, 0)$; $a = 2$;

$b = 2$.    **3.** Horizontal hyperbola; center $(0, 0)$; $a = 3$; $b = 2$.    **4.** Horizontal hyperbola; center

$(0, 0)$; $a = 2$; $b = 3$.    **5.** Vertical hyperbola; center $(0, 0)$; $a = 3$; $b = 2$.    **6.** Vertical hyperbola;

center $(0, 0)$; $a = 2$; $b = 3$.    **7.** Vertical hyperbola; center $(0, 0)$; $a = 5$; $b = 6$.    **8.** Horizontal

hyperbola; center $(0, 0)$; $a = \frac{1}{3}\sqrt{15}$; $b = \frac{1}{2}\sqrt{5}$.    **9.** Horizontal hyperbola; center $(2, -3)$; $a = 2$; $b = 4$.

**10.** Horizontal hyperbola; center $(-3, 1)$; $a = 2\sqrt{2}$; $b = \sqrt{5}$.    **11.** Vertical hyperbola; center

$(-2, -2)$; $a = \sqrt{6}$; $b = 2\sqrt{2}$.    **12.** Vertical hyperbola; center $(-1, -2)$; $a = 5$; $b = 4$.

**13.** Horizontal hyperbola; center $(2, -3)$; $a = \sqrt{2}$; $b = \sqrt{5}$.    **14.** Horizontal hyperbola; center

$(-4, -3)$; $a = 2$; $b = \sqrt{3}$.    **15.** Horizontal hyperbola; center $(-3, 1)$; $a = \sqrt{10}$; $b = \sqrt{6}$.

**16.** Horizontal hyperbola; center $(1, -2)$; $a = \frac{2}{3}$; $b = 1$.    **17.** Horizontal hyperbola;  center $(\frac{1}{3}, 1)$;

$a = \frac{1}{3}$; $b = \frac{1}{6}\sqrt{3}$.    **18.** Horizontal hyperbola; center $(0, -4)$; $a = 2\sqrt{10}$; $b = \frac{4}{3}\sqrt{30}$.    **19.** $\dfrac{y^2}{25} - \dfrac{x^2}{24} = 1$

**20.** $\dfrac{x^2}{25} - \dfrac{y^2}{11} = 1$    **21.** $\dfrac{x^2}{1} - \dfrac{y^2}{24} = 1$    **22.** $\dfrac{(y-6)^2}{4} - \dfrac{(x-4)^2}{5} = 1$    **23.** $\dfrac{x^2}{9} - \dfrac{(y+4)^2}{7} = 1$

**24.** $\dfrac{(x-2)^2}{16} - \dfrac{y^2}{3} = 1$

## Problem Set 5

**1.** (3) hyperbola; (4) hyperbola; (5) hyperbola; (6) ellipse; (7) hyperbola; (8) parabola; (9) ellipse

**2.** (10) hyperbola; (11) parabola; (12) parabola; (13) parabola; (14) ellipse; (15) hyperbola; (16) ellipse

For Problems 3-7, $\theta = 45°$; $x = \frac{1}{\sqrt{2}}(x' - y')$; $y = \frac{1}{\sqrt{2}}(x' + y')$

**8.** $\tan\theta = 2$; $x = \frac{1}{\sqrt{5}}(x' - 2y')$; $y = \frac{1}{\sqrt{5}}(2x' + y')$    **9.** $\tan\theta = \frac{1}{2}$; $x = \frac{1}{\sqrt{5}}(2x' - y')$; $y = \frac{1}{\sqrt{5}}(x' + 2y')$

For Problems 10-12, $\theta = 30°$; $x = \frac{1}{2}(\sqrt{3}x' - y')$; $y = \frac{1}{2}(x' + \sqrt{3}y')$

For Problems 13-14, $\theta = 60°$; $x = \frac{1}{2}(x' - \sqrt{3}y')$; $y = \frac{1}{2}(\sqrt{3}x' + y')$

**15.** $\theta = 45°$; $x = \frac{1}{\sqrt{2}}(x' - y')$; $y = \frac{1}{\sqrt{2}}(x' + y')$    **16.** $\tan\theta = 3$; $x = \frac{1}{\sqrt{10}}(x' - 3y')$; $y = \frac{1}{\sqrt{10}}(3x' + y')$

**17.** hyperbola; $\dfrac{x'^2}{16} - \dfrac{y'^2}{16} = 1$    **18.** hyperbola; $\dfrac{y'^2}{8} - \dfrac{x'^2}{8} = 1$    **19.** hyperbola; $\dfrac{y'^2}{2} - \dfrac{x'^2}{2} = 1$

**20.** ellipse; $\dfrac{x'^2}{9} + \dfrac{y'^2}{4} = 1$    **21.** hyperbola; $\dfrac{x'^2}{9} - \dfrac{y'^2}{4} = 1$    **22.** parabola; $(x' + 1)^2 = 4(y' + \frac{7}{10})$

**23.** ellipse; $\dfrac{x'^2}{9} + \dfrac{y'^2}{4} = 1$    **24.** hyperbola; $\dfrac{x'^2}{4} - \dfrac{y'^2}{9} = 1$    **25.** parabola; $x'^2 = 8y'$    **26.** parabola;

$x'^2 = \frac{1}{16}(y' - 4)$    **27.** parabola; $y'^2 = -12x'$    **28.** ellipse; $\dfrac{(x'-2)^2}{\frac{1}{4}} + \dfrac{(y'-1)^2}{1} = 1$

**29.** hyperbola; $\dfrac{y'^2}{4} - \dfrac{x'^2}{16} = 1$    **30.** ellipse; $\dfrac{(x' - \sqrt{10})^2}{44} + \dfrac{(y' - 2\sqrt{10})^2}{4} = 1$

## Problem Set 6

**1.** parabola; no rotation    **2.** parabola; no rotation    **3.** parabola; no rotation    **4.** circle; no
rotation    **5.** line; no rotation    **6.** hyperbola; no rotation    **7.** parabola; no rotation    **8.** point;
no rotation    **9.** ellipse; no rotation    **10.** rotated hyperbola; $\theta = 45°$    **11.** rotated parabola;
$\theta = 45°$    **12.** rotated ellipse; $\tan\theta = 2$    **13.** parabola; $(y + 1)^2 = 4(x - 5)$

**14.** parabola; $(x + 2)^2 = -12(y + 5)$    **15.** parabola; $(x - 3)^2 = \frac{2}{3}(y - 8)$    **16.** circle;

$(x + 1)^2 + (y - 2)^2 = 25$    **17.** line; $y = -\frac{25}{9}x + 25$    **18.** hyperbola; $\dfrac{(x - 3)^2}{16} - \dfrac{(y + 1)^2}{4} = 1$

**19.** parabola; $(y - 3)^2 = 4(x + 1)$    **20.** point; $(-3, 2)$    **21.** ellipse; $\dfrac{(x - 3)^2}{25} + \dfrac{(y - 4)^2}{9} = 1$

**22.** rotated hyperbola; $\theta = 45°$; $\dfrac{x'^2}{8} - \dfrac{y'^2}{8} = 1$    **23.** rotated parabola; $\theta = 45°$; $(x' + 3)^2 = 6(y' + 2)$

**24.** rotated ellipse; $\tan \theta = 2$; $\dfrac{x'^2}{9} + \dfrac{y'^2}{4} = 1$    **25.** $x^2 = -4y$: $y \geq -4$

**26.** $y = 2x - 10$: $-14 \leq y \leq -12$    **27.** $y = 2x - 8$: $-12 \leq y \leq 13.6$

**28.** $\dfrac{x^2}{64} + \dfrac{4(y + 12)^2}{1} \leq 1$: $2x - 8 \leq y \leq 2x - 10$    **29.** $\dfrac{x^2}{16} - \dfrac{3(y + 4)^2}{64} = 1$: $-12 \leq y \leq -4$

**30.** $(x + 3)^2 + (y + 15)^2 = \frac{9}{4}$    **31.** $x^2 + (y - \frac{3}{2})^2 = \frac{9}{4}$: $y \leq 2$    **32.** $\dfrac{4x^2}{1} - \dfrac{7(y - 7)^2}{25} = 1$:

$2 \leq y \leq 12$    **33.** $\dfrac{x^2}{2} + \dfrac{4(y - 12)^2}{1} = 1$: $y \geq 12$    **34.** $\dfrac{x^2}{16} + \dfrac{9(y + 4)^2}{1} = 1$: $y \geq -4$

**35.** $\dfrac{x^2}{2} + \dfrac{(y - 12)^2}{9} = 1$: $y \geq 12$

## Problem Set 7

**5.** Symmetric with respect to the origin. Extent: domain, all $x \neq 0$; range, all $y \neq 0$. Asymptotes: $x = 0$, $y = 0$. No intercepts.    **6.** Symmetric with respect to the origin. Extent: domain, all $x \neq 0$; range, all $y \neq 0$. Asymptotes: $x = 0$, $y = 0$. No intercepts.    **7.** No symmetry with respect to the $x$-axis, $y$-axis, or origin. Extent: domain, all $x \neq 0$; range, all $y \neq 1$. Asymptotes: $x = 0$, $y = 1$. Intercepts: $(0, \frac{1}{2})$, $(-1, 0)$.    **8.** No symmetry with respect to the $x$-axis, $y$-axis, or origin. Extent: domain, all $x \leq 4$; range, $y \geq 0$. Intercepts: $(0, 2)$, $(4, 0)$.    **9.** No symmetry with respect to the $x$-axis, $y$-axis, or origin. Extent: domain, all $x \neq -2$; range, all real numbers. Asymptotes: $x = -2$, $y = 2x - 3$. Intercepts: $(0, -5)$, $(-\frac{5}{2}, 0)$, $(2, 0)$.    **10.** No symmetry with respect to the $x$-axis, $y$-axis, or origin. Extent: domain, all $x \neq -2$; range, all $y \neq -7$. Intercepts: $(0, -1)$, $(\frac{1}{3}, 0)$.    **11.** No symmetry with respect to the $x$-axis, $y$-axis, or origin. Extent: domain, all $x \neq -\frac{1}{2}$; range, $y \geq -1$, $y \neq \frac{5}{4}$. Intercept: $(0, 0)$.    **12.** No symmetry with respect to the $x$-axis, $y$-axis, or origin. Extent: domain, all $x \neq -2$; range, all $y \neq -7$. Intercepts: $(0, -1)$, $(\frac{1}{3}, 0)$.

**13.** Symmetric with respect to the $x$-axis, $y$-axis, and origin. Extent: domain, $-2 \leq x \leq 2$; range $-3 \leq y \leq 3$; Intercepts: $(-2, 0)$, $(2, 0)$, $(0, 3)$, $(0, -3)$.    **14.** Symmetric with respect to the $x$-axis, $y$-axis, and origin. Extent: domain, $x \geq \sqrt{5}$ or $x \leq -\sqrt{5}$; range, all real numbers. Intercepts: $(0, \sqrt{5})$, $(0, -\sqrt{5})$.    **15.** Symmetric with respect to the origin. Extent: domain, $-\dfrac{\sqrt{26}}{2} \leq x \leq \dfrac{\sqrt{26}}{2}$; range, $-\dfrac{\sqrt{26}}{2} \leq y \leq \dfrac{\sqrt{26}}{2}$. Intercepts: $(0, \pm\frac{6}{13}\sqrt{26})$, $(\pm\frac{6}{13}\sqrt{26}, 0)$.    **16.** Symmetric with respect to the $x$-axis. Extent: domain, $x < 2$; range, all reals, $y \neq 0$. Asymptotes: $x = 2$, $y = 0$. Intercepts: $(0, 1)$, $(0, -1)$.    **17.** No symmetry with respect to the $x$-axis, $y$-axis, or origin. Extent: domain, all reals $x \neq 3$, $x \neq 1$; range, $y < -4$ or $y > 0$. Asymptotes: $x = 3$, $x = 1$, $y = 0$. Intercept: $(0, \frac{4}{3})$.

**18.** No symmetry with respect to the $x$-axis, $y$-axis, or origin. Extent: domain, $x \geq -2^{2/3}$; range, all real numbers. Asymptotes: no horizontal or vertical asymptotes. Intercepts: $(0, 0)$, $(0, -4)$.
**19.** Symmetric with respect to the $x$-axis; $y$-axis, and origin. Extent: domain, $x \leq -2$ or $-1 < x < 1$ or $x \geq 2$; range, all real numbers. Asymptotes: $x = 1$, $x = -1$. Intercepts: $(0, 0)$, $(2, 0)$, $(-2, 0)$.    **20.** Symmetric with respect to the $x$-axis. Extent: domain, $x \leq 1$ or $x > 2$; range, all reals, $y \neq 1$, $y \neq -1$. Asymptotes: $x = 2$, $y = 1$, $y \neq -1$. Intercepts: $(0, \frac{1}{2}\sqrt{2})$, $(0, -\frac{1}{2}\sqrt{2})$, $(1, 0)$.    **21.** rotated hyperbola; $\theta = 45°$    **22.** rotated hyperbola; $\theta = 45°$    **23.** rational function $y = 1/x$, which has been translated to the point $(0, 1)$    **24.** rational function $y = 1/x$, which has been translated to the point $(-2, 1)$    **25.** See answer to Problem 9 for a description.    **26.** the line $y = 3x - 1$, with a deleted point at $x = -2$    **27.** a parabola $y = x^2 - 2x$, with a deleted point at $x = -\frac{1}{2}$    **28.** a parabola $y = x^2 + 4x + 7$, with a deleted point at $x = -2$    **29.** Recognize as an ellipse: $\frac{x^2}{4} + \frac{y^2}{9} = 1$.    **30.** Recognize as a hyperbola: $\frac{y^2}{5} - \frac{x^2}{10} = 1$.    **31.** vertical ellipse with $45°$ rotation; if you carry through the rotation, the equation is $\frac{x'^2}{9} + \frac{y'^2}{4} = 1$.    **32.** See answer to Problem 16 for a description.    **33.** See answer to Problem 17 for a description.    **34.** See answer to Problem 18 for a description.    **35.** See answer to Problem 19 for a description.    **36.** See answer to Problem 20 for a description.

**Problem Set 8**

**1.** polar: $(4, \frac{\pi}{4})$; $(-4, \frac{5\pi}{4})$; rectangular: $(2\sqrt{2}, 2\sqrt{2})$    **2.** polar: $(6, \frac{\pi}{3})$; $(-6, \frac{4\pi}{3})$; rectangular: $(3, 3\sqrt{3})$
**3.** polar: $(5, \frac{2\pi}{3})$; $(-5, \frac{5\pi}{3})$; rectangular: $(-\frac{5}{2}, \frac{5}{2}\sqrt{3})$    **4.** polar: $(3, \frac{11\pi}{6})$; $(-3, \frac{5\pi}{6})$; rectangular: $(\frac{3}{2}\sqrt{3}, -\frac{3}{2})$    **5.** polar: $(\frac{3}{2}, \frac{7\pi}{6})$; $(-\frac{3}{2}, \frac{\pi}{6})$; rectangular: $(-\frac{3\sqrt{3}}{4}, -\frac{3}{4})$    **6.** polar: $(5, \frac{3\pi}{2})$; $(-5, \frac{\pi}{2})$; rectangular: $(0, -5)$    **7.** polar: $(-4, 4)$; $(4, .86)$; rectangular: $(2.61, 3.03)$    **8.** polar: $(4, 3.72)$; $(-4, .58)$; rectangular: $(-3.36, -2.18)$    **9.** polar: $(-4, \pi)$; $(4, 0)$; rectangular: $(4, 0)$    **10.** polar: $(-3, \frac{5\pi}{3})$; $(3, \frac{2\pi}{3})$; rectangular: $(-\frac{3}{2}, \frac{3}{2}\sqrt{3})$    **11.** $(5\sqrt{2}, \frac{\pi}{4})$; $(-5\sqrt{2}, \frac{5\pi}{4})$    **12.** $(2, \frac{2\pi}{3})$; $(-2, \frac{5\pi}{3})$
**13.** $(4, \frac{5\pi}{3})$; $(-4, \frac{2\pi}{3})$    **14.** $(2\sqrt{2}, \frac{5\pi}{4})$; $(-2\sqrt{2}, \frac{\pi}{4})$    **15.** $(3\sqrt{2}, \frac{7\pi}{4})$; $(-3\sqrt{2}, \frac{3\pi}{4})$    **16.** $(6\sqrt{2}, \frac{3\pi}{4})$; $(-6\sqrt{2}, \frac{7\pi}{4})$    **17.** $(2, \frac{5\pi}{6})$; $(-2, \frac{11\pi}{6})$    **18.** $(5, .64)$; $(-5, 3.79)$    **19.** $(13, 2.75)$; $(13, 5.89)$
**20.** $(7.62, 1.17)$; $(-7.62, 4.31)$    **21.** lemniscate    **22.** 4-leaved rose    **23.** 3-leaved rose    **24.** lemniscate    **25.** cardioid    **26.** cardioid    **27.** none    **28.** none (circle)    **29.** none (circle)    **30.** none (line)    **31.** cardioid    **32.** cardioid    **33.** right-cardioid    **34.** down-cardioid    **35.** up-cardioid    **36.** 4-leaved rose    **37.** 3-leaved rose; rotated 30°    **38.** 3-leaved rose    **39.** 1-leaved rose    **40.** lemniscate    **41.** lemniscate    **42.** lemniscate; rotated 45°    **43.** circle centered at the origin; notice that $r = 5 \sin \frac{\pi}{6} = 2.5$ is a constant.    **44.** circle centered at the origin; notice that $r = \cos \frac{\pi}{3} = 4.5$ is a constant.    **45.** spiral    **46.** spiral

**Problem Set 9**

*The graphs for Problems 1-16 are the same as the graphs for Problems 17-32.*

1. line; $y = -\frac{1}{2}x$    2. line; $y = 2x - 2$    3. line; $y = \frac{2}{3}x + \frac{4}{3}$    4. line; $y = -\frac{3}{5}x + \frac{9}{5}$

5. parabola; $(x + 1)^2 = 4(y - \frac{3}{4})$    6. parabola; $(x - \frac{3}{2})^2 = 9(y - \frac{23}{4})$

7. parabola; $y = x^2 + 2x + 3$    8. parabola; $y = 2x^2 - 5x + 6$    9. circle; $x^2 + y^2 = 9$

10. circle; $x^2 + y^2 = 4$    11. ellipse; $\frac{x^2}{16} + \frac{y^2}{9} = 1$    12. ellipse; $\frac{x^2}{25} + \frac{y^2}{4} = 1$

13. rotated parabola; $x^2 - 2xy + y^2 - 13x + 12y + 38 = 0$ (rotation is 45°)

14. rotated parabola; $x^2 - 2xy + y^2 + 3x - 3y + 6 = 0$ (rotation is 58°)

15. rotated parabola; $4x^2 - 4xy + y^2 + 36x - 20y + 75 = 0$ (rotation is 63.4°)

16. rotated parabola; $x^2 - 4xy + 4y^2 + 11x - 23y = 0$ (rotation is 26.6°)

**Problem Set 10**

**Sample Test 1**

1. **a.** hyperbola **b.** ellipse **c.** parabola **d.** line    2. **a.** parabola **b.** ellipse **c.** hyperbola

**d.** line    3. **a.** line **b.** hyperbola **c.** cardioid **d.** lemniscate    4. **a.** circle **b.** line **c.** rose

curve **d.** rose curve    5. $\frac{(x - 4)^2}{4} + \frac{(y - 1)^2}{3} = 1$    6. $2x - y - 1 = 0$

7. $(y - 3)^2 = 20(x - 6)$    8. $\frac{4(x + 5)^2}{25} - \frac{4(y - 4)^2}{75} = 1$    9. $\frac{(x + 5)^2}{1} - \frac{(y - 4)^2}{3} = 1$

10. vertical ellipse; $\frac{x^2}{16} + \frac{y^2}{25} = 1$    11. horizontal hyperbola; $\frac{(x + \frac{1}{2})^2}{3} - \frac{(y + \frac{1}{2})^2}{3} = 1$

12. parabola that opens up; $x^2 = y$    13. circle; $(x - 2)^2 + (y - 1)^2 = 2$    14. vertical hyperbola;

$y^2 - x^2 = 1$    15. cardioid; $r = 4(1 + \cos \theta)$    16. $x^2 - 16y + 64 = 0$    17. **a.** focus (3, 2);

center $(-3, 20)$; directrix $x + 9 = 0$    18. **a.** 45°  **b.** $\frac{1}{2}$    19. **a.** $[-\frac{\sqrt{2}}{4}, 0) \cup (0, \frac{\sqrt{2}}{4}]$

**b.** $(-\infty, \infty)$    20. **a.** origin  **b.** $x = 0$

**Sample Test 2**

1. **a.** line **b.** parabola    2. **a.** hyperbola **b.** line    3. **a.** hyperbola **b.** hyperbola

4. **a.** circle **b.** parabola    5. **a.** parabola **b.** parabola    6. **a.** cardioid **b.** lemniscate

7. $\frac{(x + 4)^2}{16} + \frac{(y + 2)^2}{15} = 1$    8. $16x^2 - 24xy + 9y^2 + 168x - 176y + 616 = 0$

9. $7x + 3y + 64 = 0$    10. $x + 3y + 5 = 0$    11. $\frac{x^2}{25} + \frac{y^2}{9} = 1$; horizontal ellipse, center (0, 0),

$a = 5; b = 3$    12. $(x + 1)^2 + (y - 1)^2 = 5$; circle, center $(-1, 1)$, radius $\sqrt{5}$

13. $(x + 3)^2 = (y + 7)$; parabola, opens up; center $(-3, -7)$; $4c = 1$; $c = \frac{1}{4}$

14. $\frac{(x + 2)^2}{3} - \frac{(y - 2)^2}{1} = 1$; horizontal hyperbola, center $(-2, 2)$; $a = \sqrt{3}, b = 1$

15. $(x - 2)^2 + (y + 2)^2 = 8$; circle; center $(2, -2)$; $r = 2\sqrt{2}$    16. $5x + 3y + 1 = 0$

**17. a.** $\theta = \frac{3\pi}{8}$   **b.** $(3, 0)$, $(-3, 0)$, $(0, \frac{3}{5}\sqrt{5})$, $(0, -\frac{3}{5}\sqrt{5})$   **18.** domain, all reals; range,

$-\frac{1}{2} \leq y < 0$ or $0 < y \leq \frac{1}{2}$   **19. a.** origin   **b.** $y = 0$   **20. a.** center $(2, -1)$; $\epsilon = \frac{1}{2}$; major

axis 4; minor axis $2\sqrt{3}$   **b.** center $(5, -4)$; $\epsilon = \frac{\sqrt{34}}{3}$; transverse axis 6; conjugate axis 10

# INDEX